T0174386

Digital Audio Theory

Digital Audio Theory: A Practical Guide bridges the fundamental concepts and equations of digital audio with their real-world implementation in an accessible introduction, with dozens of programming examples and projects.

Starting with digital audio conversion, then segueing into filtering, and finally real-time spectral processing, *Digital Audio Theory* introduces the uninitiated reader to signal processing principles and techniques used in audio effects and virtual instruments that are found in digital audio workstations. Every chapter includes programming snippets for the reader to hear, explore, and experiment with digital audio concepts. Practical projects challenge the reader, providing hands-on experience in designing real-time audio effects, building FIR and IIR filters, applying noise reduction and feedback control, measuring impulse responses, software synthesis, and much more.

Music technologists, recording engineers, and students of these fields will welcome Bennett's approach, which targets readers with a background in music, sound, and recording. This guide is suitable for all levels of knowledge in mathematics, signals and systems, and linear circuits. Code for the programming examples and accompanying videos made by the author can be found on the companion website, DigitalAudioTheory.com.

Christopher L. Bennett is a Professor in the Music Engineering Technology program at the University of Miami, Frost School of Music. He conducts research, teaches, and publishes in the fields of digital audio, audio programming, transducers, acoustics, psychoacoustics, and medical acoustics.

Digital Audio Theory

A Practical Guide

Christopher L. Bennett

Routledge
Taylor & Francis Group

LONDON AND NEW YORK

First published 2021
by Routledge
2 Park Square, Milton Park, Abingdon, Oxon OX14 4RN

and by Routledge
52 Vanderbilt Avenue, New York, NY 10017

Routledge is an imprint of the Taylor & Francis Group, an informa business

British Library Cataloguing-in-Publication Data
A catalogue record for this book is available from the British Library

Library of Congress Cataloging-in-Publication Data
Names: Bennett, Christopher L., author.
Title: Digital audio theory: a practical guide / Christopher L. Bennett.
Description: Abingdon, Oxon; New York, NY: Routledge, 2021. |
Includes bibliographical references and index.
Subjects: LCSH: Sound—Recording and reproducing—Digital techniques. |
Signal processing—Digital techniques. | Digital communications.
Classification: LCC TK7881.4 .B47 2021 (print) | LCC TK7881.4 (ebook) | DDC 621.389/3—dc23
LC record available at https://lccn.loc.gov/2020031085
LC ebook record available at https://lccn.loc.gov/2020031086

ISBN: 978-0-367-27655-3 (hbk)
ISBN: 978-0-367-27653-9 (pbk)
ISBN: 978-0-429-29714-4 (ebk)

Typeset in Minion
by codeMantra

Visit the companion website: DigitalAudioTheory.com

Dedicated to Alicia, Claire & Jack

Contents

3 Sampling — 33

4 Aliasing and reconstruction — 47

5 Quantization — 65

9 z-Domain 125

10 IIR filters 147

Abbreviations

AC	alternating current
ADC	analog to digital converter
AM	amplitude modulation
BPF	band-pass filter
DAC	digital to analog converter
DAW	digital audio workstation
dB	decibel
DC	direct current
dec	decade
DFT	discrete Fourier transform
DSP	digital signal processing
DUT	device under test
ESS	exponential swept-sine
EQ	equalization
FFT	fast Fourier transform
FIR	finite impulse response
FOIL	first, outside, inside, last
FM	frequency modulation
grd	group delay
HF	high frequencies
HPF	high-pass filter
Hz	Hertz (cycles per second)
IDFT	inverse discrete Fourier transform
IFFT	inverse fast Fourier transform
IIR	infinite impulse response
IR	impulse response
LF	low frequencies
LPF	low-pass filter

LSB	least significant bit
MLS	maximum length sequence
MSB	most significant bit
oct	octave
PDF	probability density function
PZ	pole/zero
rad	radian
RMS	root mean square
RLC	resistor, indictor, capacitor
RPDF	rectangular PDF
s	second
SER	signal to error ratio
SNR	signal to noise ratio
TPDF	triangular PDF
ZOH	zero order hold

Abbreviations

Variables

$a_{0...n}$	feedforward coefficients
$b_{0...n}$	feedback coefficients
C	capacitance
d	delta train
D	DFT of delta train
e	Euler's constant
e_q	quantization error
f	frequency (in Hz)
f_N	Nyquist rate
f_S	sampling rate
h	filter impulse response
H	filter transfer function
i, j	$\sqrt{-1}$
k	delay amount, DFT index
K	length of convolution output
L	sweep rate
n	sample index
M	order of denominator
N	filter order, DFT order
N_{bits}	bit depth
R	resistance
q	quantization level
T_S	sample period
w	filter state variable
W_N	DFT twiddle factor
x	input sequence
X	DFT of input sequence
y	output sequence

Y	DFT of output sequence
z	complex variable
z_o	zero locations on z-plane
z_x	pole locations on z-plane
∂	Dirac delta
Δ	small change
η	zero-mean, random noise
ν	dither
ω	frequency (in rad/s)
$\mathcal{F}$	Fourier transform
$\lvert\ldots\rvert$	magnitude operator
$\angle$	phase or angle operator

1

Introduction

If you've had prior experience with a Digital Audio Workstation (DAW), then you already have some idea of how audio flows from the sound source, such as a microphone or synthesizer into the DAW via an audio interface for processing, then back out for reproduction over loudspeaker or headphones. This encompasses the capture of analog audio and its conversion to digital audio, the processing of digital audio with filters and effects, and finally the conversion of digital audio for reproduction as analog sound. In *Digital Audio Theory*, the theoretical underpinnings of this signal chain will be examined, with an emphasis on practically implementing the theory in a signal processing environment such as MATLAB® or Octave.

The digital audio signal flow to capture, process, and reproduce audio begins and ends with the converters; namely, the analog to digital converter (ADC) and the digital to analog converter (DAC). These converters are an interface between digital audio and analog representation of audio, normally voltage. Within the digital domain, typical operations of digital audio often include storage to disk, processing with a digital effect, or analysis of frequency content. The mathematical framework and practical implementation of this process will be the purview of *Digital Audio Theory* (Figure 1.1).

1.1 Describing audio signals

When recording analog sound, it is useful to classify the captured audio as either desired or undesired (let's call the latter "noise"). This classification depends on the type of sound we hope to capture – typically we might think of an instrumentalist, vocalist, or speech signal, but the numbers of categories are nearly endless, they

1

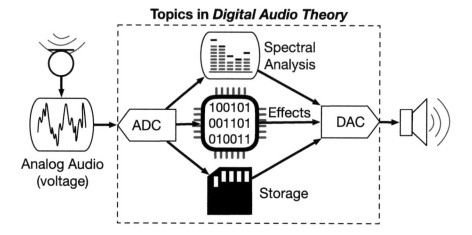

Figure 1.1

Overview of topics covered in this text, which include analog/digital conversion, linear effects (such as filters), spectral analysis, and processing.

could be ecological (e.g., urban soundscape or wildlife sounds), physiological (e.g., lung or cardiovascular sounds), among many others. However, what could be considered our desired signal in one context, could be considered noise in another. For example, environmental sounds at a sporting event are often intentionally mixed in with the broadcast to give a sense of immersion, but these same environmental sounds may be considered noise when capturing film dialog. In addition to the ambient soundscape captured by a microphone, we could also add other types of noise, including electrical (e.g., ground hum or hiss) and mechanical (e.g., vibrations of the microphone). Each of these can further be classified by their duration; *transient* sounds are short duration while *steady-state* sounds ongoing or periodic.

1.1.1 Measuring audio levels

With acoustic sound, we measure its level in units of pressure, the Pascal (Pa), which is simply force over an area (N/m²). When sound travels through air, we are not measuring the actual pressure of the air, but rather the pressure fluctuation around static pressure, which is around 101,325 Pa at sea level. Sound Pressure Level (SPL) fluctuations about static pressure that would typically be captured range anywhere from less than 1 mPa to as great as 10 Pa. The level of an acoustic audio signal can be reported as its absolute peak amplitude (known as peak SPL), or the range from its lowest trough to its highest peak (peak-to-peak SPL), or as its average value, typically reported as its root-mean-square (RMS) value. Unless otherwise specified, an SPL value can be assumed to be the RMS level, given by:

$$x_{RMS} = \sqrt{\frac{1}{N} \sum_{n=1}^{N} x_n^2} \qquad (1.1)$$

This equation tells us to take every value in our audio signal, x_n, and square it. Then sum all of those values together and divide by the total number of values, N, giving the average of the squared values. Finally, we take the square root of the mean of the squared values to obtain the RMS.

Without diving into psychoacoustics, or the study of the perception of sound, it can be noted that our ears perceive sound logarithmically. This applies to both SPL as well as frequency. For example, a doubling of frequency corresponds to an octave jump. To the human ear, an octave interval sounds the same, irrespective of the starting frequency. For example, the interval from 100 Hz to 200 Hz (a 100 Hz range) sounds perceptually similar to the interval from 200 Hz to 400 Hz (a 200 Hz range). For this reason, the ear is said to hear frequencies on a logarithmic base-2 scale, or $\log_2$. For SPLs, the ear also hears logarithmically, but we use base-10 instead, or $\log_{10}$. The unit that audio is typically reported in is a decibel (dB_{SPL}), defined as

$$dB_{SPL}\left(x_{RMS}\right) = 20 \cdot \log_{10}\left(\frac{x_{RMS}}{20\ \mu Pa}\right) \tag{1.2}$$

Here, the signal, x_{RMS}, is converted to a logarithmic scale, with a reference of 20 μPa, the quietest SPL perceivable by the human ear. It is not uncommon to see dB_{SPL} reported simply as "dB", but this is incorrect since a dB is strictly a ratio between any two values, while a dB_{SPL} is a ratio between a SPL and 20 μPa. Another common dB unit in audio is $dB_{Full\text{-}Scale}$, or simple dB_{FS}. "Full Scale" refers to the dB ratio between an audio level and the maximum representable level by the system, therefore the unit dB_{FS} could be thought of as the dB below Full Scale. In a digital audio system, the largest representable value is fixed – we can assign this level any arbitrary value, but 1.0 is typical. If we measure, in the same digital audio system, a signal with an RMS level of 0.1, then its dB_{FS} can be calculated as

$$dB_{FS}\left(0.1\right) = 20 \cdot \log_{10}\left(\frac{0.1}{1.0}\right) = -20\ dB_{FS} \tag{1.3}$$

1.1.2 Pro-audio versus Consumer audio levels

You may also be familiar with the units dBu and dBv. Just like with dB_{SPL}, the letters "u" and "v" indicate a specific reference value. The reference for dBv is 1 Volt (V) – this is the reference that is used for consumer audio. The consumer audio standard level, which is −10 dBv, corresponds to an RMS voltage level of $10^{\frac{-10}{20}} \cdot 1.0 = 0.316$ V. On the other hand, pro audio, which is reported in dBu, uses a reference voltage of 0.775 V. This voltage represents the level at which 1 milliWatt (mW) of power is achieved across a 600 Ohm (Ω) load, which was a historical standard impedance for audio equipment. (Power, voltage, and resistance are related by $P = \frac{V^2}{R}$, therefore $\frac{0.775^2}{600} = 1$ mW). Pro audio equipment operates at +4 dBu, which corresponds to an RMS voltage level of $10^{\frac{+4}{20}} \cdot 0.775 = 1.228$ V. The actual dB difference between pro and consumer audio equipment is $20 \cdot \log_{10}\left(\dfrac{1.228}{0.316}\right) = 11.8\ dB.$

1.1.3 Dynamic range

If you are a musician, you are already familiar with dynamics, ranging from *piano* to *forte*. A dynamic range is simply the difference between the loudest and quietest sound. In audio, we can use dynamic range to describe an acoustic source (e.g., the dynamic range of an orchestra is about 85 dB) or of audio equipment (e.g., the dynamic range of a microphone could be 100 dB). Importantly, dynamic range tells us nothing about the upper or lower limits, only the difference between these two extremes. When selecting a microphone, it is important to consider the limits of the dynamic range of a microphone with respect to those of the source that is being recorded.

1.1.4 Signal to noise ratio (SNR)

Another type of range that we should consider when describing audio signals is that between a signal's level and the *noise floor* of the recording. The noise floor is the noise that exists in an audio capture system even in the absence of an acoustic source. Electrical noise within the equipment itself can contribute to the noise floor, as can even thermal agitation of air molecules, which can be detected by a sensitive microphone. The noise *floor* is so-called because it represents the absolute lowest level of capture for audio. The difference between a signal's level and the noise floor is known as SNR, as shown in Figure 1.2. In the digital domain, we will encounter a type of noise that is specifically created through the conversion process, which also contributes to the noise floor.

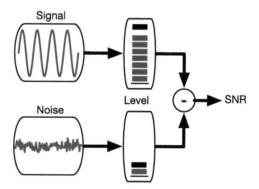

Figure 1.2

The SNR is given by the difference (in dB) of the levels of the signal and noise, respectively.

There are some rules of thumb to remember about SNR. First, if SNR is negative, then the noise is a higher level than the signal (uh-oh!), and SNR is positive when the signal level is higher (and 0 dB if they are exactly the same level). Additionally, if the signal level is twice the noise level, we see that the SNR is $20 \cdot \log_{10} 2 = +6.02$ dB. Furthermore, we can add or subtract 6.02 dB for every doubling or halving of the SNR, respectively. For example, a ratio factor of 4× implies a +12.04 dB SNR, or inversely, a ratio of 0.5× implies a −6.02 dB SNR. The same concept applies to factors of 10×, which correspond to steps of 20 dB. Other useful rules of thumb are, factors of 3× ⇨10 dB and factors of 30× ⇨30 dB. These can be also be combined to quickly convert between dB and scale factor without the use of a calculator; for example, 16 dB = 10 dB + 6 dB. Since the addition of log values is a multiplication of the operands, we can estimate the linear scale factor as 3·2 = 6×. Just to verify, we can use a calculator to see that $10^{\frac{16}{20}} = 6.3×$ – pretty close to our estimation using some simple rules of thumb.

1.1.5 Frequency and bandwidth

In addition to an audio signal's level, we also characterize it by its frequency and phase, measured in Hz and radians (rad), respectively. An audio signal that is simply a single frequency, can be represented as

$$x(t) = A \cdot \sin(2\pi ft + \phi) \tag{1.4}$$

Here, the signal level is set by the scalar value A, and the frequency (or the number of complete cycles made by the sinusoid) is represented by f, and the starting point of the sinusoid is given by the phase, Φ. The reason for the 2π is to indicate that every $t = 1$ s, the sinusoid is making f complete rotations or cycles.

We can visualize this process for both a sin and a cos function (recall that $\sin\left(2\pi ft + \frac{\pi}{2}\right) = \cos(2\pi ft)$). Consider the circle shown in Figure 1.3 (a), starting at the '0' marking (3 o'clock position) with markers every $\frac{\pi}{8}$ rad in the counterclockwise direction. From each of these markers, if we draw a vertical line down until it intersects the horizontal axis, we see the *projection* of the circle at that specific angle onto this axis. Next, plot the angles on a horizontal line, as shown in Figure 1.3 (b), then on the vertical axis plot the value of the projection (onto its horizontal axis) for that given angle. It can be seen that this process traces out a cosine function in Figure 1.3(b). Through a similar process, we can trace out a sine function by looking at the projection of the circle at each angle onto the vertical axis. For this reason, when plotting a circle sometimes the horizontal axis is called the cosine axis, while the vertical axis is called the sine axis.

But audio signals typically comprise more than one frequency. It is often useful to indicate the relative levels of each of the constituent frequencies. In this type of graph, known as the spectrum, frequency is laid out on the horizontal axis and level is on the vertical axis (see Figure 1.4). For example, on a parametric equalizer (EQ), there are a few knobs corresponding to frequency. Those frequencies can be

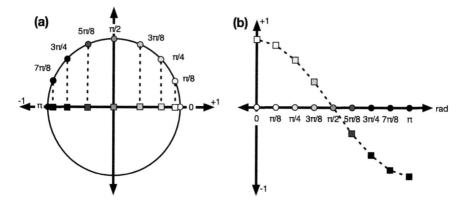

Figure 1.3

ingles from 0 to π in steps of π/8 are shown in circle markers in both (a) and (b). Projections of these positions along the circle onto the horizontal axis are shown in square markers in (a). These values are reproduced in (b) for each angle. Projecting onto the horizontal axis traces out half a cycle of a cosine. If we continued on from π to 2π, then we would see the rest of the cosine traced out.

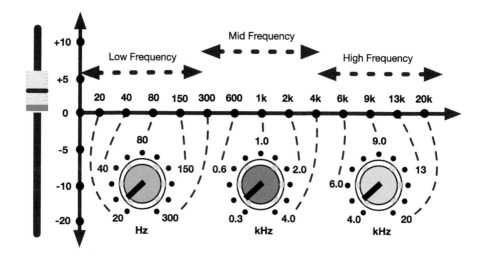

Figure 1.4

By mapping frequencies from 20 to 20 kHz to the horizontal axis, and mapping levels (in dB) to the vertical axis, we establish a way to visualize the spectrum of an audio signal. For every frequency on the horizontal axis, a value is drawn that corresponds to the SPL of that frequency component.

mapped to the linear frequency axis. We can consider frequencies from the low end of hearing (around 20 Hz) up to around 300 Hz to be "low frequencies", those from 300 Hz to about 4.0 kHz as "mid frequencies", and finally from 4.0 kHz to 20.0 kHz as "high frequencies". For any frequency component in our signal, we mark a value, the height of which indicates the SPL of that particular frequency.

Just like we described the difference between the highest and lowest SPLs as the dynamic range, we can also describe the difference between the highest and lowest frequencies as the *bandwidth*. And just like the dynamic range, a bandwidth can apply to a source or a system. For example, the bandwidth of the human auditory system is about 20 kHz, and the bandwidth of the notes on an 88-key piano is 4,158.5 Hz, spanning 27.5 Hz (A0) to 4,186 Hz (C8).

Bandwidth can also be represented as a number of octaves or decades, rather than a frequency range, and this has to do with the logarithmic spacing of *pitch* or the perception of frequency. An octave interval has a constantly changing bandwidth, depending on the starting frequency. For example, the bandwidth from G4 to G5 is 392 Hz, while the bandwidth from G5 to G6 is 784 Hz. To our ear, this sounds like a fixed interval, but in reality, the bandwidth of an octave increases with frequency. For this reason, frequencies (especially in audio) are plotted not on a linear axis but on a logarithmic axis, like the horizontal axis in Figure 1.4.

1.1.6 Characterizing noise

Noise is often characterized based on its spectral properties, and in audio there are two common types. The first type is known as *white noise*, so called because it contains equal amounts of all frequencies, similar to white light. Therefore, a spectral plot of white noise will be flat, as shown in Figure 1.5, since all frequencies contain the same energy. However, to the ear, which hears logarithmically, as we ascend in frequency, each subsequent octave band actually contains twice as much energy as the previous octave. Therefore, perceptually, white noise sounds weighted to the high frequencies, or bright. The second type of noise is known as *pink noise*, is inversely proportional to frequency, and contains equal amounts of energy per octave. For this reason, pink noise sounds more natural to the ear, and in fact, pink noise is more commonly found in nature. The energy in pink noise decreases at a clip of 3 dB per octave (or equivalently, 10 dB per decade).

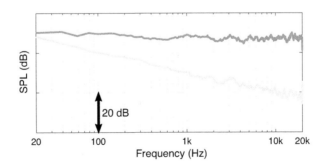

Figure 1.5

White noise (dark gray) contains equal energy across all frequencies, while pink noise (light gray) contains equal energy across all octaves.

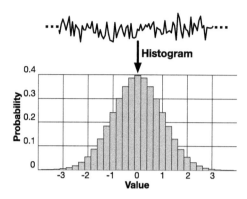

Figure 1.6

Zero-mean Gaussian white noise (top) plotted as a histogram (bottom). The most likely values are centered around zero, with decreasing probability as value increases.

1.1.7 Probability density functions and histograms

With the spectrum plot, the level of each frequency (horizontal axis) is plotted on the vertical axis. With a probability density function (PDF), the number of occurrences of each level is plotted on the vertical axis, with the signal level on the horizontal axis. A histogram answers the question, "Considering all of the values in a given signal, how often does each value appear, or in other words what is the likelihood of any particular value?" For audio, zero will typically be the most probable value, exhibiting decreasing probability with increasing value (in both the positive and negative directions). This kind of value distribution is often described as being *zero-mean Gaussian*, since the probability roughly follows a Gaussian curve. A way of visualizing a PDF for a sequence of values is with the histogram – a *binned* version of a PDF. A histogram still represents the distribution by dividing the entire range of values into a series of intervals and counting the number of values that fall within each interval. A histogram of zero-mean Gaussian white noise is shown in Figure 1.6.

1.2 Digital audio basics

In the analog domain, audio is almost always represented as a voltage level that is continuous in both time and level. In other words, "zooming in" to the signal produces an increasingly precise measure of the level for an infinitesimally decreasing time interval. To represent these signals on a computer (a word I will use as a synecdoche for all digital audio processors), the signal must first be digitized by converting the continuous voltage into a sequence of numbers with finite precision – that is to say, both the temporal as well as dynamic aspects of the signal will be discretized.

Sampling and *quantization* are the processes by which a continuous signal is digitized for subsequent digital signal processing. To achieve digitization, first, a

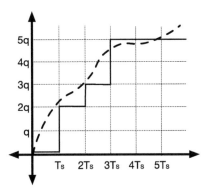

Figure 1.7

A signal that is continuous in time and level (dashed) is sampled with a period of T_S with a quantization step size of q (solid).

time interval will be established in which to "check" the value of the analog signal (a process known as *sampling*). That value must then be truncated to a finite precision in order to be stored on a computer (a process known as *quantization*). This time interval is assigned the variable T_S (the subscript 's' stands for 'sampling'), while its inverse is the *sampling frequency*, $f_S = 1/T_S$. A high precision clock ensures even spacing of T_S from one sample to the next.

The values occurring every T_S must then be quantized for digital representation at 16, 20, 24, 32, or 64 bits, a number known as the *bit-depth*, which dictates the dynamic range and resolution capabilities – the larger the bit-depth, the better the dynamic range and resolution. Audio distribution formats are typically in 16-bits, which can represent 65,535 levels of resolution, q. ADCs will typically capture at 16-bits, 20-bits (1,048,575 levels), or 24-bits (16,777,215 levels), while audio processing should occur at a minimum of 24-bits, but increasingly DAWs are moving to 32-bit (4,294,967,296 levels) and 64-bit (1.844674407 × 10^{19} levels). Raising the bit-depth improves representation of the "true" analog signal and decreases digital artifacts during processing. A representation of a sampled (solid line) analog signal (dashed line) is shown in Figure 1.7.

1.3 Describing audio systems

When it comes to audio, we are often concerned with how a system or device affects the signal – in other words, how are the frequency response, gain, envelope, or other properties impacted. An audio device could be anything from a transducer (microphone or speaker), to an amplifier, or an outboard processor (e.g., compressor or equalizer). In digital audio, the systems comprise digital states (such as a memory element) rather than analog states (such as a capacitor). For both digital and analog audio systems, we want to characterize them along several criteria, just like we do with audio signals.

1.3.1 Continuous and discrete systems

Just like a signal can be characterized as continuous or discrete, a system or device can also be continuous or discrete. In the case of a discrete system, the properties change at specific intervals of time (you can probably already guess that these time intervals are the sampling interval, T_S), while the properties of a continuous system change constantly over time. Analog audio systems typically operate on continuous time and continuously varying levels of voltages, while digital audio systems have discrete units of time (dictated by the sample rate) as well as quantized representation of level (governed by bit-depth).

1.3.2 Linear and non-linear systems

Another way that an audio system is characterized is by its *linearity*. A linear system is one that obeys the *scalar* and *additive* rules. The scalar rule of linear systems states that if a signal is multiplied (scaled) by some value, then the output of the system will also be scaled by that same value. Consider a linear audio system, H, that we pass in an audio signal, x, generating output y.

$$y = H(x) \tag{1.5}$$

For system that follows the scalar rule, if we then scale x by a, its output will also be scaled by a.

$$a \cdot y = a \cdot H(x) = H(a \cdot x) \tag{1.6}$$

Similarly, the additive rule states that adding two signals together at the input produces the same output as adding their individual outputs together. Consider the following two signals, x_1 and x_2 passed through system H generating outputs y_1 and y_2

$$y_1 = H(x_1) \tag{1.7a}$$

$$y_2 = H(x_2) \tag{1.7b}$$

If H adheres to additive rules of linearity, then the following is also true:

$$y_1 + y_2 = H(x_1) + H(x_2) = H(x_1 + x_2) \tag{1.8}$$

If a system follows both of these rules, then we can characterize it as linear. An input-output (I/O) graph is useful for indicating the linearity of a system. An I/O graph indicates the input level on the horizontal axis and the output level on the vertical axis, along with a *characteristic curve*. The characteristic curve maps the input level to the output level. In linear systems, the characteristic curve is a line (as it stands to reason), and in non-linear systems the curve may have a variety of shapes. A series of I/O graphs are shown in Figure 1.8. Figure 1.8 (a) demonstrates the scalar property for a linear system, while Figure 1.8 (b) shows the same inputs with a non-linear system.

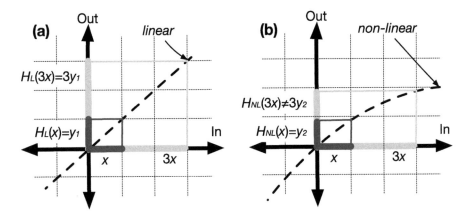

Figure 1.8

(a) An input, x, with amplitude shown on the horizontal axis, produces an output, y_1, in a linear system, H_L, with its output level indicated on the vertical axis. When the input is scaled, the output is also scaled by the same amount. (b) The same input is shown for a saturating non-linear system, H_{NL}. When the input to a non-linear system is scaled, its output, y_2, is not scaled by the same amount.

1.3.3 Temporal properties

The temporal aspects of an audio system involve how the system behaves at different time points, and whether the system is capable of operating in *real-time*. In audio, real-time systems are often utilized, for example, an effects processor that is being used on an instrument during a live performance. For a system to be real-time simply means that it can generate outputs at least as quickly as it is receiving inputs. You may have experienced a situation when working with a DAW when the computer cannot process playback and halts – in this scenario, the computer cannot maintain real-time processing.

A pre-requisite for real-time processing is that the system is *causal* – when the system's output only depends on the instantaneous and previous inputs and outputs and does not depend on future inputs. Since a non-causal system contains future inputs or outputs, it is not practically realizable as a real-time system. Note that this can be circumvented by storing inputs and treating slightly delayed signal as the current input, then the system has access to "future" parts of the signal equal to the amount of delay.

Another temporal property of an audio system is its *time variance*. A classic example of a time-variant system is a tube amplifier when it is first turned on. Irrespective of the input to the tube amp, there may be no output for the first several seconds. And as the cathode warms (freeing electrons), the sound character of the amplifier may evolve for another ten minutes or more. If a system is time *invariant* (also called "shift invariant"), then it will produce identical outputs at different time points, if given the same input and will sound the same from the moment it is turned on until it is turned off.

1.3.4 Impulse response

An audio system that is causal, linear, and time invariant (sometimes abbreviated LTI or LSI) can be completely described simply by its *impulse response* (IR). The IR of any system, analog or digital, is its output to a very specific input known as an impulse or a delta, δ. The δ is zero valued except for one moment in time when it rises to a very high value and falls immediately back to zero. While the δ itself is very short in duration, the IR, which is designated by the variable h, can be much longer (even infinitely long). The IR is the output of the system, H, when inputted with a delta or impulse, δ.

$$h = H(\delta) \tag{1.9}$$

For an acoustic system, such as a room response, the impulse can be generated by a pistol firing blanks, a balloon pop, or a wooden "clapper" that slaps two flat planks of wood together. In a digital system, the impulse is generated by setting the value of all samples of a sequence of samples to zero with the exception of a single sample that is set to one. The IR tells us how much "ringing" is present in a system and the relative levels of frequencies.

The IR can also indicate whether the system is *stable*. Stability insures that as long as the input values stays within some bounds, that all future outputs of the system will also stay within some bounds. An example of an unstable audio system would be an echo effect where each subsequent echo was *louder* than the previous echo. As time passed, the output would get louder and louder, eventually approaching infinite loudness. An audio system is considered stable if its IR satisfies the following:

$$\sum_{t=0}^{\infty} |h_t| < \infty \tag{1.10}$$

Since a finite length IR could never fail this criterion, we only concern ourselves with infinite length IRs as potentially being unstable. But generally, as long as the output is fading toward zero, then stability will likely be achieved. In later chapters, we will revisit stability for different filter types.

1.3.5 Frequency response

We will want to characterize how an audio system *filters*, or affects, the frequency response of an input audio signal; in other words, which frequencies get amplified and which get attenuated. To describe a filter, we must first define some terminology for it. The primary characteristics of a filter are its *pass-* and *stop-bands*. The pass-band is the frequency region that is unaffected by the filter, while the stop-band is the frequency region that is attenuated by the filter. In between the pass- and stop-bands is the *cutoff frequency*, sometimes called the

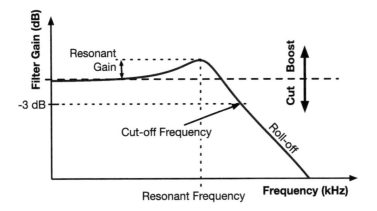

Figure 1.9

Here a filter is shown with the major properties of its frequency response labeled, including the cutoff frequency (which occurs at the −3dB point), the resonance (with resonant gain and resonant frequency), and the roll-off. The pass-band is the frequency region before the cutoff, and the stop-band is beyond it. The pass- and stop-bands will be defined differently for different filter types.

corner or critical frequency. The cutoff frequency (or frequencies) is where the filter reaches −3 dB (or half-power). Around the cutoff frequency, there may be a *resonance*, or a boost that has a resonant gain and resonant frequency. Finally, the *roll-off* is the rate at which the attenuation increases beyond the cutoff frequency; typically, this would be reported in dB/octave or dB/decade. All of these concepts are shown in Figure 1.9.

Filters can also be classified by the location of their pass- and stop-bands. In Figure 1.9 a *low-pass filter* (LPF) is shown – so called because low frequencies are allowed to pass through the filter, while high frequencies are attenuated. The reverse of this would be a *high-pass filter* (HPF). Other filter types target specific bands; one such filter is the *band-pass filter* (BPF), which only allows frequencies in a certain region to pass, while attenuating both higher and lower frequency regions. The opposite of this is a *band-reject filter* (BRF), which rejects a specific frequency region, while allowing lower and higher frequencies to pass. Related is the *notch filter*, which is a highly tuned BRF that attenuates a narrow frequency region. If the notch is repeated at a regular frequency interval throughout the entire spectrum, then the filter type is known as a *comb filter*. In another class of filters, a particular frequency region is boosted rather than attenuated. If a low or high band is boosted, then the filter is known as a *shelving filter*, which can be a high-shelf (HSF) or a low-shelf (LSF). And if a narrow-band is boosted, then this filter type is known as a *peaking filter* or *resonator*. A final type of filter is the *all-pass filter* (APF), which does not affect the relative level of any frequency, but only modifies the phase. Each of these filter types is shown in Figure 1.10.

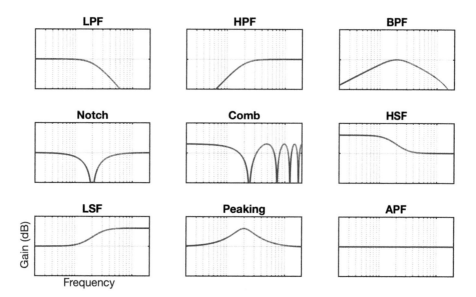

Figure 1.10

Several classes of filters are shown. The first five have pass- and stop-bands, and they are low-pass filter, high-pass filter, band-pass filter, notch filter, and comb filter. The next three have pass and boost bands, and they are the high-shelf filter, low-shelf filter, and peaking filter. The last is the all-pass filter, which does not affect the level of any frequency, only the phase.

1.4 Further reading

This book assumes some pre-requisite knowledge in the areas of music, math, engineering, and programming. A list of recommended texts is listed below to supplement a few of the important background topics.

As a companion to this text, I recommend *Hack Audio* by Eric Tarr. This book provides an introduction to MATLAB® programming, specifically for musical effects. It is handy for musicians and audio engineers interested in computer programming.

Eric Tarr. *Hack Audio: An Introduction to Computer Programming and Digital Signal Processing in MATLAB* (1st ed.). New York, NY: Routledge, 2019.

Physics and Music by Harvey and Donald White provides an introduction to the science behind sound and music for students without a strong background in physics and mathematics. In particular, the first nine chapters cover general principles of sound, musical scales, the primary ways in which sound can be generated and are highly related to the topics in *Digital Audio Theory*.

Harvey White and **Donald White**. *Physics and Music: The Science of Musical Sound* (Reprint). Mineola, NY: Dover Publications, 2014.

As a primer on Trigonometry and Pre-Calculus topics, I recommend *Algebra and Trigonometry* by Michael Sullivan. It covers graphing on a cartesian plane, solving polynomial equations, trigonometry, logarithms, polar coordinates, sequences, and probability, which are all pre-requisite knowledge for *Digital Audio Theory*.

Michael Sullivan. *Algebra and Trigonometry* (10th ed.). London, UK: Pearson, 2016.

While knowledge of signals and systems, linear circuit theory, and Laplace transformations are <u>not</u> a pre-requisite for *Digital Audio Theory*, the interested reader can find these topics, along with additional information on analog filters, *s*-domain, and continuous time transfer functions in:

Larry Paarmann. *Design and Analysis of Analog Filters: A Signal Processing Perspective*. Dordrecht, NL: Kluwer Academic Publishers, 2003.

If *Digital Audio Theory* is material that you are already familiar with, and you would like to pursue more in-depth application of digital audio to music effects and synthesis, then I can highly recommend the following two books by Will Pirkle that cover digital audio effects and synthesis:

Will Pirkle. *Designing Audio Effect Plugins in C++: For AAX, AU, and VST3 with DSP Theory* (2nd ed.). London, UK: Routledge, 2019.

Will Pirkle. *Designing Software Synthesizer Plug-Ins in C++: For RackAFX, VST3, and Audio Units* (1st ed.). London, UK: Routledge, 2014.

1.5 Challenges

1. Which level has a higher voltage: consumer line level at +6 dBv or pro line level at −6 dBu?
2. Resonant frequency is to cutoff frequency as resonant gain is to _____?
3. What type of filter should be used to remove a DC (0 Hz) bias?
4. What type of filter should be used to eliminate a 60 Hz ground noise that contains integer harmonics?
5. Write the function that describes a sinusoid with a frequency of 100 Hz and a phase offset of π rad.
6. Calculate the RMS of a sinusoid with amplitude ±1.0
7. What textbook contains additional MATLAB® programming examples of audio effects?

1.6 Project – audio playback

These Projects are designed for the mathematics-oriented computing environments MATLAB® (The Mathworks, Inc.) and Octave (a GNU, open source). For the most part, the syntax between these two platforms is identical, with some small differences that will be indicated in code comments throughout. This first project serves as an introduction to signal generation and plotting and should be accessible to those with some prior programming or scripting experience. For the completely uninitiated, some additional resources are listed in the previous section.

Generating sinusoidal signals

    i. Create two sinusoids of differing frequencies, amplitudes (use amplitudes < 0.5), and phase. Using a 3 × 1 subplot, graph each signal and the sum of the two signals. Add a title as well as labels to the horizontal and vertical axes.

   ii. Play the output of the combined signal. Can you hear both frequencies?

Wave files

    i. Read a stereo.wav file from disk (< 30 seconds). Some public domain samples are available for download from DigitalAudioTheory.com

   ii. Using a 2 × 1 subplot, graph the left and right channels.

  iii. Play the.wav file at higher and lower sample rates. Make note of how the sound changes.

Mixing audio

    i. Generate a 440 Hz tone of the same length as the wave file

   ii. Add the tone to the wave file into a new variable

  iii. Plot and play the output. To avoid clipping, scale the sound by its largest value

  iv. Save this new file to disk as an audio filetype

Hints for Lab 1

Here is a list of commands to use:

    i. Generating sinusoids – `sin` and `cos`

   ii. Plotting – `plot` and `subplot`

  iii. Formatting a plot – `xlabel`, `ylabel`, `title`, and `axis`

  iv. Normalizing – `max` and `abs`

   v. Audio commands – `audioread`, `audiowrite`, `audioplayer`

2

Complex vectors and phasors

This chapter provides a segue into sampling theorem and frequency transformations by way of complex vectors and phasors. Sampling (Chapter 3) is one half of the digital conversion process (the other half is quantization, Chapter 5), while frequency transformations are important for analysis (Chapter 12) and spectral processing (Chapter 13). Suffice it to say that complex vectors and phasors are foundational to understanding the concepts of digital audio theory.

In this chapter, the complex plane and the unit circle will be introduced as a means of representing complex vectors. This is a powerful tool for digital audio that will be utilized to explore the meaning behind Euler's formula, explain the phenomenon of beat frequencies, and allow the reader to develop their own amplitude and frequency modulated sounds in the project at the end of this chapter. We will start with a review of complex number representation, basic properties, and operations. Next, we will look at plotting complex vectors on the complex plane, followed by an introduction to complex conjugates and phasors.

2.1 Complex number representation and operations

Before starting on complex numbers, let's pause and reflect briefly on negative numbers; at one time, these were considered by some to "darken the very whole doctrines of the equations" (English Mathematician, Francis Maseres, 1759). But

Table 2.1 Comparison of Negative and Complex Numbers. Note the Use of the $Re\{...\}$ and $Im\{...\}$ Operators – These Indicate That We Are Extracting Just the Real and Imaginary Parts, Respectively, of a Complex Number

Property	Negative Numbers	Complex Numbers
Answers the question	What is < 0?	What is $\sqrt{-1}$?
Implication	Same magnitude in the <u>opposite</u> direction	The same magnitude in a <u>rotated</u> direction
Impact on number line	Adds a new half to the left of 0	Adds a vertical axis to the number line
Multiplication cycle	$x, -x, x, -x, ...$	$x, j \cdot x, -x, -j \cdot x, x,...$
Magnitude	$\|x\| = \sqrt{(-x)^2}$	$\|x\| = \sqrt{Re\{x\}^2 + Im\{x\}^2}$

negative numbers had utility in practice and in nature; as an example, removing the air contents from a sealed container results in a negative pressure (we might say a vacuum) with respect to atmospheric pressure, and therefore, is best expressed as a negative number. If we take the natural number line that starts at zero and add to it negative numbers, then a new direction (to the left) is created, with zero representing the middle rather than the beginning.

Complex numbers, which contain real and imaginary (denoted in this text with the variable j) parts, are also useful for representing natural phenomena. For example, in electrical engineering, the amount of impedance that a reactive component (such as an inductor or capacitor) imposes on a circuit depends on the frequency of the current running through it. This phenomenon is perfectly modeled with complex numbers. In a way, there is a duality between negative and complex numbers, shown in Table 2.1. Negative numbers doubled the number line by creating a new direction that starts at 0 and goes to the left. Complex numbers have a similar impact on the number line, which is again divided, but with a vertical axis that extends from negative (downward) to positive (upwards), as shown in Figure 2.1.

By adding an imaginary axis to our number line, we can change the direction of a vector on the horizontal number line from positive to negative (or vice versa) *without altering the* magnitude. This is done simply by imagining the value being "rotated" and thus only altering its angle rather than its magnitude.

2.1.1 Unit circle

Rotations within the complex plane (or the plane formed by the *real* and *imaginary* axes) are not limited to steps of j. If we imagine a unit vector (meaning magnitude of 1) and rotate it continuously to $+j$ then -1 then $-j$ and back to $+1$, the circle that this traces out is known as the *unit circle*. The unit circle sits on the cartesian plane, but is a convenient way to graph complex vectors. The unit circle is centered at $(0,0)$ and extends up to ±1 on the real axis and $\pm j$ on the imaginary axis. Unlike a clock, which begins at the 12:00 position and rotates clockwise, on the unit circle, an angle of 0 rad is at $+1$ on the real axis (or the 3:00 position) and rotates counter-clockwise.

2. Complex vectors and phasors

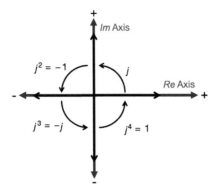

Figure 2.1

The real vector with value +1 can be rotated to –1 without ever changing its magnitude of 1 by adding an imaginary axis to the complex plane. Multiplication by j causes the vector to rotate into the imaginary axis. Multiplication by j again rotates the vector to the negative half of the real axis, then again to the negative half of the imaginary axis, and then finally back to a real positive value.

The angular space that the unit circle (or any circle for that matter) encompasses is 2π radians (rad), meaning that –1 on the unit circle is at π rad, and +j and –j are at $\pi/2$ rad and $-\pi/2$ rad (or equivalently $3\pi/2$ rad), respectively. Complex vectors that aren't entirely on either of the two axes can be represented by the amount that they project onto each axis – this is known as *rectangular form*. The amount of projection onto the real axis is determined by the cosine of the angle between the real axis and the vector, while the projection onto the imaginary axis is given by the sine of this same angle (Figure 2.2).

The complement to rectangular form is *polar form*, in which the complex vector is represented by its *magnitude* and *angle*. The magnitude, r, indicates the amplitude

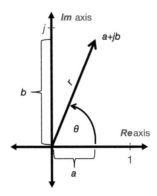

Figure 2.2

The vector, x, is represented by its amount of projection onto the Real axis $a = Re\{x\} = r \cdot \cos(\theta)$ and the Imaginary axis $b = Im\{x\} = r \cdot \sin(\theta)$ and is represented in rectangular form as $x = a + j \cdot b$

of the vector, while the angle, θ, indicates the amount of rotation, and is expressed as $r \cdot e^{j\theta}$. From rectangular form, the magnitude and angle can be computed from the following:

$$r = \sqrt{\text{Re}\{x\}^2 + \text{Im}\{x\}^2} \qquad (2.1a)$$

$$\theta = \text{atan}\left(\frac{\text{Im}\{x\}}{\text{Re}\{x\}}\right) \qquad (2.1b)$$

Polar form is normally expressed in terms of the base of the natural logarithm, $e \approx 2.71828\ldots$, using Euler's formula, which relates e to the sine and cosine functions:

$$r \cdot e^{j\theta} = r \cdot \cos(\theta) + j \cdot r \cdot \sin(\theta) \qquad (2.2)$$

This equation tells us that for an imaginary sine and cosine function each with magnitude r, the magnitude of their complex sum will always be equal to r, irrespective of the angle θ. When r is set equal to one and the angle θ is varied from 0 to 2π, the unit circle is traced. Using Euler's formula, the cosine can be written as a sum of complex exponentials, which is equal to the projection of a vector onto the real axis:

$$\cos(\theta) = \frac{e^{j\theta} + e^{-j\theta}}{2} = \frac{a + jb + a - jb}{2} = a \qquad (2.3)$$

While the sine can be written as

$$\sin(\theta) = \frac{e^{j\theta} - e^{-j\theta}}{2} = \frac{a + jb - a + jb}{2} = j \cdot b \qquad (2.4)$$

For this reason, often the real axis is referred to as the "cosine" axis – the projection of a complex vector onto the real axis is the cosine of the angle of that vector. This remains similarly true for the sine and the imaginary axis. The sum of two complex vectors is shown in Figure 2.3, illustrating the cancellation of imaginary components and geometrically proving the Euler's formula for a cosine.

2.1.2 Example: convert from polar to rectangular form

Problem: Given the complex vector $x = 0.5 e^{\frac{j\pi}{6}}$, find its equivalent rectangular representation.

Solution: The projections onto the real and imaginary axes must be determined, using the cosine and sine, respectively.

$$a = \text{Re}\{x\} = 0.5 \cos\left(\frac{\pi}{6}\right) = \frac{\sqrt{3}}{4}$$

2. Complex vectors and phasors

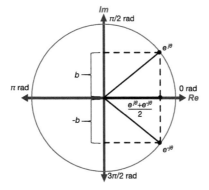

Figure 2.3

For some angle, θ, the cosine can be visualized geometrically as a mean of two complex sinusoids whose real components sum together and whose imaginary components cancel out, resulting in a real value.

$$b = \operatorname{Im}\{x\} = 0.5 \sin\left(\frac{\pi}{6}\right) = \frac{1}{4}$$

$$x = a + j \cdot b = \frac{\sqrt{3}}{4} + j \cdot \frac{1}{4}$$

Double Check: To verify the answer, we can determine the magnitude and angle from our proposed solution.

$$r = \sqrt{a^2 + b^2} = \sqrt{\left(\frac{1}{4}\right)^2 + \left(\frac{\sqrt{3}}{4}\right)^2} = \sqrt{\frac{4}{16}} = \frac{1}{2}$$

$$\theta = \tan^{-1}\left(\frac{b}{a}\right) = \tan^{-1}\left(\frac{1/4}{\sqrt{3}/4}\right) = \frac{\pi}{6}$$

2.1.3 Programming example: plotting complex vectors

We can visualize this the complex vector above on the complex plane (Figure 2.4).

```
theta = pi/6;      % angle
r = 1/2;           % magnitude
x=r*exp(j*theta);  % complex vector
N=10;              % number of data points

% trace out a line from origin to r
% at a constant angle
polarplot(theta*ones(N,1), linspace(0,r,N)) %Matlab
polar(theta*ones(N,1), linspace(0,r,N))      %Octave
title('x=0.5e^{\pi/6}')
% The following are Matlab only:
```

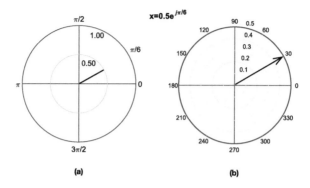

(a)   (b)

Figure 2.4

A vector can be generated on the complex plane. (a) Using the **polar** plot, the radius and units of the axes can be set. (b) Using **compass**, the axes are fixed and the units are in degrees.

```
pax = gca;
pax.ThetaAxisUnits = 'radians';
rlim([0 1]) % Matlab only

% or use compass, which plots a vector on the complex plane
figure;
compass(x);
```

Additionally, we can utilize commands that are native to MATLAB® or Octave to extract real and imaginary components from polar form, as well as calculate magnitude and angle from rectangular form.

```
% get projection onto real axis Re{x}
a = r * cos(theta)

% get projection onto imaginary axis Im{x}
b = r * sin(theta)

% use native commands to get the same result
real(x)
imag(x)

% double check with pythagorous
Mag = sqrt(a^2 + b^2)
Ph = atan(b/a)

% native commands
abs(x)
angle(x)
```

From the code examples above, it can be seen that the **real**() and **imag**() commands can be used to extract the real and imaginary components of a complex number, respectively. Additionally, the **abs**() and **angle**() commands indicate the magnitude and angle of a complex number.

2.1.4 Complex mathematical operations

Addition and Subtraction: When adding or subtracting two complex numbers together, the real and imaginary components are summed (or subtracted) separately. If the number is in polar form, it is easier to first convert to rectangular form. For example, the subtraction of two complex vectors is given by:

$$(a+jb)-(c+jd)=(a-c)+j(b-d) \tag{2.5}$$

Multiplication and Division: In rectangular form, multiplication uses the First-Outside-Inside-Last (FOIL) method, shown below:

$$(a+jb)\cdot(c+jd)=ac+jbc+ajd+j^2bd=(ac-bd)+j(bc+ad) \tag{2.6}$$

Multiplication and division are much easier in polar form. For multiplication, the magnitudes are multiplied and the angles are added, while for division, the magnitudes are divided and the angles are subtracted, as shown here:

$$r_1e^{j\theta_1}\cdot r_2e^{j\theta_2}=r_1r_2e^{j(\theta_1+\theta_2)} \tag{2.7a}$$

$$\frac{r_1e^{j\theta_1}}{r_2e^{j\theta_2}}=\frac{r_1}{r_2}e^{j(\theta_1-\theta_2)} \tag{2.7b}$$

2.2 Complex conjugates

A complex vector has a "partner", known as its *complex conjugate*, which is denoted by '*' such that x^* is the conjugate of x. Conjugates mirror each other about the real axis; this implies that they share the same magnitude and same real component, but have opposite angles, and therefore opposite imaginary components. In polar form, the complex conjugate of $x=re^{j\theta}$ is $x^*=re^{-j\theta}$ while in rectangular form, the complex conjugate of $x=a+j\cdot b$ is $x^*=a-j\cdot b$. Note that changing the sign of the angle is identical to changing the sign of the imaginary component. One useful property of conjugate pairs is that they have the same magnitude, $|x^*|=|x|$. If they are multiplied together, their product is a real valued number equal to the magnitude of the vector squared, such that $x\cdot x^*=|x|^2$. We can prove this relationship:

$$x\cdot x^*=(a+jb)\cdot(a-jb) \tag{2.8a}$$

$$=a^2-j\cdot ab+j\cdot ab-j^2b^2 \tag{2.8b}$$

$$= a^2 + b^2 = |x|^2 \qquad (2.8c)$$

It is easy to visualize a complex conjugate on the complex plane, as shown in Figure 2.3, where a vector and its complex conjugate are shown. Note that they have equal magnitudes and equal real components, but opposite angles and opposite imaginary components.

2.3 Phasors

A *phasor* is simply a vector that is rotating. Pop Quiz: What is the frequency of $\cos(\pi/2)$? Actually, this is a trick question, because the cosine of a scalar is just a scalar; in other words, it has no frequency since the value of the angle is not changing. But if we multiply the angle by a value that is constantly and consistently increasing (e.g., time, t), then we cause the angle to increase and therefore the vector to rotate, as shown in Figure 2.5. Consider the following complex phasor

$$x = 0.5e^{\frac{j\pi}{6}t} = 0.5 \cdot \cos\left(\frac{\pi}{6}t\right) + j \cdot 0.5 \cdot \sin\left(\frac{\pi}{6}t\right) \qquad (2.9)$$

Here, x has a frequency of $\omega = \dfrac{\pi}{6} rad/s$, or equivalently in cycles per second a frequency of $f = \dfrac{\omega}{2\pi} = \dfrac{1}{12}$ Hz, which can equivalently be expressed as:

$$x = 0.5e^{j2\pi\frac{1}{12}t} = 0.5 \cdot \cos\left(2\pi\frac{1}{12}t\right) + j \cdot 0.5 \cdot \sin\left(2\pi\frac{1}{12}t\right) \qquad (2.10)$$

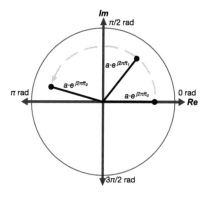

Figure 2.5

At three arbitrary points in time (t_0, t_1, t_2) the angle of the vector increases, resulting in rotation of the vector. This is known as a phasor.

2. Complex vectors and phasors

In other words, it takes 12s for this phasor to make one complete rotation from 0 rad to 2π rad. Note that a complex phasor, represented by a single vector, contains both a real as well as an imaginary component.

While a single rotating phasor is complex (with both real and imaginary parts), a time-domain audio signal is purely real, so we need to develop a representation for purely real signals with no imaginary components. Starting from a real signal, x_R, we can write it as:

$$x_R = \cos(2\pi ft) \tag{2.11}$$

Next, we can add and subtract the same amount of an imaginary component, without changing the fact that it is still purely real:

$$x_R = \cos(2\pi ft) + j0.5\sin(2\pi ft) - j0.5\sin(2\pi ft) \tag{2.12}$$

Next, if we break up the cosine portion into two halves, we can express x_R as:

$$x_R = 0.5\cos(2\pi ft) + 0.5\cos(2\pi ft) + j0.5\sin(2\pi ft) - j0.5\sin(2\pi ft) \tag{2.13}$$

Finally, by rearranging terms, it can be seen that a real signal is really a sum of two complex conjugates, in which the imaginary components cancel out:

$$x_R = \underbrace{\left[0.5\cos(2\pi ft) + j0.5\sin(2\pi ft)\right]}_{0.5e^{j2\pi ft}} + \underbrace{\left[0.5\cos(2\pi ft) - j0.5\sin(2\pi ft)\right]}_{0.5e^{-j2\pi ft}} \tag{2.14}$$

This is simply restating Euler's formula, but with phasors instead of vectors – the addition of a complex phasor with its conjugate produces a real-valued sum. This is a pretty striking statement at first – to produce a real digital audio sinusoid, we require two complex phasors rotating at the same velocity, but in opposite directions, each with half of the amplitude. Therefore, to plot a real signal (such as an audio signal) with a frequency > 0 Hz on the complex plane, **two** components must be used so that each holds half of the magnitude (see Figure 2.6) at plus and minus the frequency associated with f, whereby:

$$a \cdot \cos(2\pi ft) = \frac{a}{2} \cdot \left(e^{j2\pi ft} + e^{-j2\pi ft}\right) \tag{2.15}$$

2.3.1 Programming example: rotating a vector

In this example, we will create a complex phasor by rotating a vector. This is achieved by incrementing the angle of a vector using a time variable, t. The vector, v, has a magnitude of $a = 0.5$, and a frequency of $f = 1$ Hz. A 'for-loop' is utilized here to iterate time from $t_0 = 0$ s in steps of $t_{step} = 1/12$ s up to $t_{max} = 0.5$ sec. At each time point, the vector is plotted on polar coordinates on the same figure (using the **hold on** command). A different RGB triplet, comprising red, green, and blue

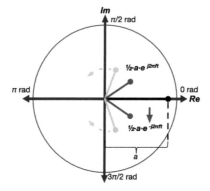

Figure 2.6

A purely real phasor, $x_R = a \cdot \cos(2\pi ft)$, has two parts in polar form, which form a conjugate pair, each with half of the magnitude of the signal. The phasor moving clockwise (denoted by the arrow) has opposite phase of the phasor moving counter-clockwise. The real parts sum constructively on the horizontal axis, while the imaginary parts, on the vertical axis, cancel. At time $t = 0$ (black), there is no conjugate and the full magnitude, a, is on the Real axis. At subsequent time points (gray, light gray), the magnitude splits, $\frac{1}{2}a$, between conjugate phasors moving in opposite directions. As the phasors rotate, their sum moves back and forth from +1 to –1 on the real axis in sinusoidal motion.

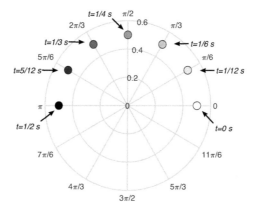

Figure 2.7

Here the phasor $x = 0.5 \cdot e^{j2\pi 1 t}$ is plotted for times $t = 0$ s to $t = 0.5$ s in equal time steps of $t_{step} = 1/12$ sec. Since the frequency is f = 1 Hz, it traverses half of the Unit Circle (0 rad to π rad) in half of a second.

values, respectively, is assigned to the marker face color for each time point. When the RGB channels are the same, a grayscale color is generated. By calculating the marker color with time, a different grayscale color accompanies each time point. For improved visualization, the marker size is set to 20 points and the marker edge is set to black ('**k**'). Finally, the polar angle units are set to radians. The end result is shown in Figure 2.7.

2. Complex vectors and phasors

```
a=0.5;             % magnitude
t_step=1/12;       % sec
t_max=0.5;         % sec
f = 1;             % Hz
vv = [];           % will hold array of vectors
for t = 0:t_step:t_max
    ang = 2*pi*f*t;
    v = a*exp(j*ang);
    vv = [vv v];          % concatenate array of vectors
    gs = 1-t/t_max;  % RGB triplet value
    % polarplot is Matlab only for Octave, comment
    % the next line and use the polar() outside the for-loop
    polarplot(v, 'Marker', 'o', 'MarkerSize', 20, ...
        'MarkerFaceColor', [gs gs gs], 'MarkerEdgeColor', 'k');
    hold on;
end

pax = gca;                    % Matlab only
pax.ThetaAxisUnits = 'radians';    % Matlab only

polar(0:t_step*2*pi:t_max*2*pi, abs(vv), 'o') % Octave only
```

2.4 Beat frequencies

Musicians may already be familiar with the phenomenon of beat frequencies. Beating occurs when two sounds of nearly identical frequencies are played simultaneously. This often occurs when tuning an instrument. When two tones are nearly the same, a modulation of the amplitude is heard, and the closer together the two frequencies, the slower the modulation becomes. Consider two sinusoids: an in-tune one, x, and a mistuned one x', with the same amplitude a and slightly different frequencies, f and $f + \Delta$. In the example below, the sinusoids will be complex. The math for real signals is exactly the same in principle, but with some additional algebra. Consider the heard signal, y, that comprises two tones of slightly differing frequencies:

$$y = \underbrace{a \cdot e^{j2\pi ft}}_{x} + \underbrace{a \cdot e^{j2\pi(f+\Delta)t}}_{x'} \tag{2.16}$$

By multiplying and dividing the first term by $e^{j2\pi\frac{\Delta}{2}t}$, and then by factoring out $e^{j2\pi\frac{\Delta}{2}t}$ from the second term, the equation can be rewritten as

$$y = a \cdot e^{-j2\pi\frac{\Delta}{2}t} \cdot \underline{e^{j2\pi\left(f+\frac{\Delta}{2}\right)t}} + a \cdot e^{j2\pi\frac{\Delta}{2}t} \cdot \underline{e^{j2\pi\left(f+\frac{\Delta}{2}\right)t}} \tag{2.17}$$

Next, the above-underlined term, which is a phasor with a frequency that is an average of the in-tune and mis-tuned signals, can be factored out. This technique of factoring out the average frequency was described by the famous physicist Richard

Feynman for characterizing a beating effect that occurs with light lasers of very nearly identical frequencies [1]. Adopting his technique, we obtain the following, which can be viewed as a carrier and a modulator:

$$y = \underbrace{a \cdot e^{j2\pi\left(f+\frac{\Delta}{2}\right)t}}_{\text{average of } x \text{ and } x'} \cdot \underbrace{\left(e^{-j2\pi\frac{\Delta}{2}t} + e^{j2\pi\frac{\Delta}{2}t} \right)}_{2\cos\left(2\pi\frac{\Delta}{2}t\right)} \tag{2.18}$$

Using Euler's formula, it can be seen that what remains inside the parentheses is a cosine with a frequency of $\Delta/2$ and an amplitude of 2 – this is the *beat* frequency, or the amplitude modulation that is applied to the carrier signal. By inspection, it can be seen that the closer the two tones become in frequency, the slower the beating becomes. This effect is easily heard when a guitar is being tuned – as the tuning pin is slowly turned, the rate of beating can be heard changing as well. It is possible to visualize this effect if we look at the magnitude of y, which is:

$$|y| = a \cdot 2\cos\left(2\pi\frac{\Delta}{2}t \right) \tag{2.19}$$

The magnitude of the carrier a is a real scalar, so it can be plotted on the horizontal axis. The modulator comprises two conjugate phasors that sum to a real, but changing, value that is also on the horizontal axis. The sum of a with the effect of the beating is illustrated in Figure 2.8. As the beat phasors (each with half the magnitude of the beat frequency) rotate, the effect of beating sometimes scale the amplitude greater than a, and sometimes (for example, as shown in Figure 2.8) scale it less than a. This geometrically shows the rising and falling of the amplitude with frequency $\Delta/2$.

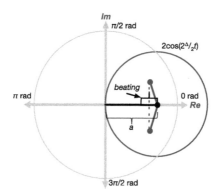

Figure 2.8

The magnitude, a, of the carrier frequency is a real, constant scalar (black vector). This magnitude is modulated by a slowly changing beat frequency, that comprises two conjugate phasors (dark gray) that sum to a real value (recall, a cosine is the sum of two conjugate phasors). The beat amplitude is a sum of their projection onto the real axis. The product of this changing value with a results in the beating effect that is heard when two tones with nearly identical frequencies are summed.

2. Complex vectors and phasors

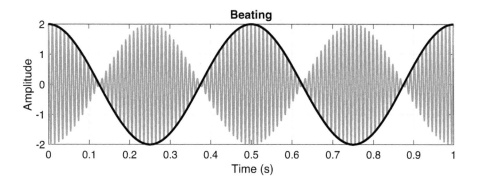

Figure 2.9

Plotting the sum of two cosines with frequencies f and $f+\Delta$ results in a carrier with a frequency of $f+\Delta/2$ Hz (you can prove this to yourself by counting the number of peaks that occur in 1 sec). The carrier is modulated by a beat frequency of $\Delta/2$ Hz. However, the ear hears this as a beating of Δ Hz since it cannot distinguish between the positive and negative phases of the modulation.

2.4.1 Programming example: beat frequencies

Here, we will programmatically show that a sum of two cosines with nearly identical frequencies is equal to a carrier of their average frequency with a modulator of half the difference frequency. A 100 Hz cosine, x_1, and a 104 Hz cosine, x_2, are summed and plotted, as in Figure 2.9. This is the same as a carrier with a frequency of 102 Hz ([104 + 100]/2 Hz) that is modulated by a cosine of 2 Hz ([104 – 100]/2 Hz). This 4 Hz beating can be heard here:

```
t=0:0.0001:1;          % time vector
x1=cos(2*pi*100*t);    % tone 1 (100 hz)
x2=cos(2*pi*104*t);    % tone 2 (104 hz)
plot(t, x1+x2)         % combination

% plotting a tone at 102 Hz with a beat frequency of 2 Hz
beat=2.*cos(2*pi*2*t);
x3=cos(2*pi*102*t)*beat;

hold on;
plot(t, x3)
plot(t, beat)
```

2.5 Challenges

1. Given the following two complex vectors:

$$x_1 = -2 + j \cdot 3$$

$$x_2 = 0.5 - j$$

Evaluate the following:

$$y_1 = x_1 + x_2$$

$$y_2 = \frac{x_1}{x_2^*}$$

2. For the following complex vector:

$$x_3 = 0.75e^{\frac{j\pi}{2}}$$

Calculate:
   a. $\ln(x_3)$
   b. $x \cdot x^*$
   c. $\mathrm{Re}\{x\} + \mathrm{Im}\{x\}$

3. For the following complex phasor:

$$z = e^{-j2\pi ft}$$

Prove the following:
   a. z is a circle.
      Hint 1: The magnitude of a circle is always 1
      Hint 2: Use Euler's formula.

   b. $\sin(j2\pi ft) = \frac{1}{2j}(z^* - z)$
   c. $z \cdot z^* = |z|^2$
   d. $\dfrac{z^*}{z} = e^{2 \cdot j2\pi ft}$
   e. $z + z^* = 2\mathrm{Re}\{z\}$
   f. $|z_1 z_2| = |z_1| \cdot |z_2|$

2.6 Project – AM and FM synthesis

In this project, you will synthesize complex tones using Amplitude Modulation (AM) and Frequency Modulation (FM). The concept of AM was already covered in section **2.4 Beat Frequencies**. AM occurs when a carrier, for example $a_c \cdot \cos(2\pi f_c t)$, is multiplied by another signal with different amplitude and frequency, $a_m \cdot \cos(2\pi f_m t)$. The product of these two signals is the AM signal, given by $a_c \cdot (1 + a_m \cdot \cos(2\pi f_m t)) \cdot \cos(2\pi f_c t)$. An AM effect is often simply described by its modulation "depth" (a_m) and modulation frequency, f_m. While AM utilizes a modulator to vary the *amplitude* of a carrier over time, FM utilizes the modulator to vary the *frequency* of the carrier over time and is given by $a_c \cdot \cos(2\pi f_c t + a_m \cos(2\pi f_m t))$.

Amplitude modulation synthesis and Tremolo

  i. Write a script that uses AM to synthesize a sound with a 700 Hz carrier and a 300 Hz modulator.
 ii. Plot a few waveforms of the AM signal. Listen to the sound using **audioplayer()** and write it to disk with **audiowrite()**.
iii. Repeat steps i. and ii. with a 50 Hz modulator.
 iv. Write a script to produce a *tremolo* effect using an audio wav file as the carrier signal and a sine wave with a subsonic frequency, that is, <20 Hz, as the modulator. Listen to the output and save the sound to disk.

Problem 2: frequency modulation

  i. Repeat steps i. through iii. above using frequency modulation.

Bibliography

[1]   Feynman, R.P., *The Feynman Lectures on Physics (Vol. 1)*, Addison-Wesley, Reading, MA, Ch. 48-1.

3

Sampling

Why sample? The reason is quite simply this – while there is an infinite number of possible time points (and corresponding signal values), a digital system can only store a finite number of values in memory. So, what *is* sampling? Sampling is one half of the digital audio conversion process, which also includes reconstruction or the process of converting digital audio back to analog audio for reproduction. If improperly designed or executed, a digital sampler can be destructive (or lossy) and even introduce distortions to the signal. But if the sampler meets certain specifications, then these artifacts can be reduced or avoided altogether.

Sampling is the *isochronic* process of capturing a signal level. "Isochronic" means equally separated in time – in digital signal processing, this is also often described as *discrete-time*, which is also assumed to be isochronic. In a non-audio sense, an example of sampling could be as simple as stepping outside each morning to gauge the temperature – in this case, the sampling period (or the amount of time that passes between subsequent observations) is one day. However, the weather changes more rapidly than once per day; while it may be warm in the morning, a cold front could suddenly pass over in the afternoon. That is to say, the sampling period of once per day is not sufficient to accurately characterize the temperature throughout the entire day. This is not too dissimilar (except for all the ways in which it is) from sampling audio – we must ask ourselves the same question, "Is the signal being sampled frequently enough such that it is changing predictably between observations?"

If a signal becomes predictable over a short period of time, then there is no need to check its value during that predictable period – this is the principal concept behind sampling and reconstruction! Or to rephrase this, we could pose the question:

for some given sample period, T_S, how accurately is a continuous signal being represented by a discrete sequence of values? This is the fundamental question that motivates this chapter, and it is a question that will be addressed from a theoretical perspective. In addition to how accurately a signal is being represented, this chapter will also cover the types of signal operators that are used for sampled sequences, namely accumulators and time-shift operators.

3.1 Phasor representation on the complex plane

Referring back to Figure 2.5 in the previous chapter, it can be seen that a phasor is not easily representable on the complex plane; in that figure, the position of the phasor changes with time, t. But this is only because the complex plane only represents an angle, from 0 to 2π rad, a representation that will be called "Phase Space". If we were to instead plot a phasor with the time component <u>divided out</u>, then what remains is a static vector (for example, $a \cdot e^{j2\pi f}$), which can be plotted on the complex plane, since the angle is now static. But now instead of plotting a magnitude and angle (rad), we are actually plotting a magnitude and *frequency* (rad/s). This representation will be called "Frequency Space", as shown in Figure 3.1.

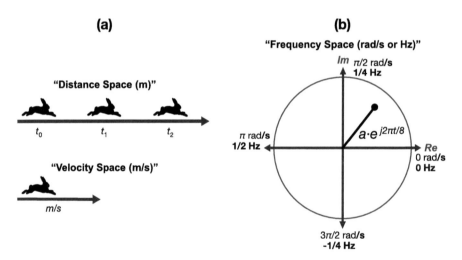

Figure 3.1

(a) If we imagine a hare running along a line representing distance, then its position is dependent on time, t. But if we divide distance by time, then we can convert the graph to a representation of velocity, and we know that the hare covers the vector length shown every second. (b) Using the same tactic of dividing by time, we can plot this phasor $\left(a \cdot e^{\frac{j2\pi t}{8}} \right)$ on the complex plane, knowing that it will rotate $2\pi/8$ rad every second, which is to say that it will take 8 s for it to go all the way around the unit circle, so it has a frequency of $f = 1/8$ Hz. We will call this "Frequency Space" – the differences of labeling compared to "Phase Space" are bolded.

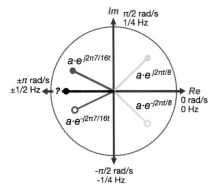

Figure 3.2

Five phasors are shown, two that are rotating in a positive (counter-clockwise) direction, indicated by filled circles, and two that are rotating in a negative (clockwise) direction, indicated by open circles. The light gray phasors are at 1/8 Hz, while the dark gray phasors are at 7/16 Hz. But the black phasor is ambiguous – while it is clear that it is rotating at 1/2 Hz, it is not clear whether it is rotating in a positive or negative direction.

By using "Frequency Space", a phasor is easily representable, as a vector in which the angle between it and the horizontal axis is **equal to the angle "step size" every second.** By converting to Frequency Space coordinate system, we must also relabel the polar grid in terms of frequency instead of the angle. For example, units of rad become rad/s, and if divided by 2π then the unit can be expressed in Hz; note in Figure 3.1 that the negative Imaginary axis is labeled as $-1/4$ Hz. This is simply because a phasor that appears in the bottom half of the complex plane is rotating clockwise instead of counter-clockwise.

While we can consider a phasor in the top half of the complex plane as rotating counter-clockwise and in the bottom half as rotating clockwise, at a certain point some confusion sets in. For example, in Figure 3.2 there are phasors on increasing positive and negative frequency. But at exactly 1/2 Hz, it is unclear whether the phasor is rotating counter-clockwise (positive) or clockwise (negative). In actuality, there is no way to tell them apart, and for this reason the phasors that can be represented here must be limited to the bandwidth of up to (but not including) 1/2 Hz. This band-limit is, of course, not very useful for representing frequencies in the audible range!

3.2 Nyquist frequency

For the phasor $a \cdot e^{j2\pi\left(\frac{1}{2}\right)t}$ on the complex plane (in Frequency Space, as shown in black on Figure 3.2), it is impossible to know whether it is rotating counter-clockwise or clockwise, therefore we must set an exclusive limit of 1/2 Hz for representing a phasor in Frequency Space on the complex plane. However, there is a simple trick to allow us to represent frequencies greater than 1/2 Hz, which is to decrease the time window we are examining. At the moment, we are considering the phase "step size" that occurs *every second*. But we can just change our observation window in order to consider frequencies >1/2 Hz. Consider the phasor x_S where the time component, t, is scaled by some factor T_S.

$$x_S = 0.5e^{j2\pi\frac{1}{2}(t \cdot T_S)} \tag{3.1}$$

If we let the time scale factor $T_S = 0.001$ s, then the highest frequency that we can observe gets scaled by a factor of 1,000. In other words, the highest frequency we can represent within Frequency Space is a signal that rotates up to π rad (half a cycle) in 1/1000th of a second. Therefore, in 1 s that signal rotates 1000π rad, or 500 Hz. The smaller we make T_S, the higher frequency we can capture, according to the rule:

$$f_N = \frac{1}{2T_S} = \frac{f_S}{2} \tag{3.2}$$

where f_N is the "Nyquist frequency", named after the American engineer Harry Nyquist. Simply stated, f_N is equal to half the frequency given by $1/T_S$, which is known as the sampling frequency, f_S. Therefore, as f_S is increased (or as T_S is decreased), then the highest frequency that can be unambiguously represented, f_N, also increases. Figure 3.3 shows the same 1/2 Hz phasor, but with $T_S = 0.001$ s ($f_S = 1,000$ Hz). Now, the 1/2 Hz phasor is very close to the horizontal axis – the position of the phasor from 0 rad to π rad can be thought of as a percentage from 0 Hz to f_N Hz.

The process of adjusting T_S to represent higher and higher frequencies on the complex plane is functionally identical to the process of sampling in the time domain. For a real valued audio signal, we can envision a sampling function that repeatedly captures the signal value at a fixed time interval. At these fixed intervals in time, we capture the signal value, and in the interim, we ignore the signal value.

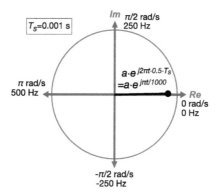

Figure 3.3

If we set T_S to 0.001 s, then to represent a 0.5 Hz phasor, we determine how much phase it covers in T_S s. In this case, only 0.001π rad are covered in 0.001 s, which is why it is *barely* above the real axis, but when we divide the angle by $2\pi T_S$, we see that the frequency is still 0.5 Hz.

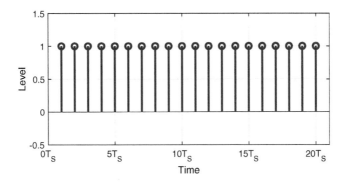

Figure 3.4

The sampling function is a sequence of ones that "capture" the signal value, with zeros in between, separated by time intervals of T_S.

This is conceptually similar to multiplying the signal by a '1' at each sampling interval, and multiplying it by a '0' in between sampling intervals. This series of 1's occurring in steps of T_S is known as the sampling function, which resembles a *train* of one-valued impulses, with zeros in between, as shown in Figure 3.4.

Each line segment is a shifted delta, where a delta is defined as:

$$\delta(n) = \begin{cases} 1, t = 0 \\ 0, t \neq 0 \end{cases} \tag{3.3}$$

To construct the "delta train" comprising many deltas, as shown in Figure 3.4, we must utilize time-shift operators. The delta train will be a combination of many individual deltas shifted to different integer multiples of T_S.

3.3 Time shift operators

The basis of linear digital audio processing is the time-shift operator. Implementing many types of digital audio effects (including many time-based, spectral, and modulation effects), involves current as well as previous inputs and outputs. To retain these non-concurrent values and utilize them at different time points, we will make use of time-shift operators known as *delays* and *advances*. We need to make use of these time-shift operators to construct the delta train shown in Figure 3.4, which will become our sampling function.

By way of example, let's suppose you are meant to take a digital audio course in the third year of your degree program. We will consider this third year to be $n = 0$, and furthermore, we can characterize the second year as $n = -1$, and the fourth year as $n = +1$ (relative to the third year). Therefore, if you were to take the digital audio course in your second year ($n = -1$), you would be *advanced* by one year, and if you had to take it in your fourth year ($n = 1$), you would be *delayed* by one year. The same concept applies to signals – if we advance a signal, it arrives earlier in time, and if we delay it, it arrives later in time. If we wanted to time-shift the delta function, it would be expressed like this:

$$\delta(n \cdot T_s + k \cdot T_s) \tag{3.4}$$

where n is the index, k is the shift amount, and the time period, T_S. In the previous example, T_S is one year. Let's say that a student is a bit clever and she took the course a year early. For the delta function to evaluate as 1 when $n = -1$, we must set $k = 1$ (advance). This can be demonstrated by iterating $n = [-2, -1, 0, 1]$, such that $nT_S = [-2T_S, -T_S, 0, T_S]$ whereby

$$\delta(-2 \cdot T_S + 1 \cdot T_s) = \delta(-1 \cdot T_s) = 0 \tag{3.5a}$$

$$\delta(-1 \cdot T_S + 1 \cdot T_s) = \delta(0) = 1 \tag{3.5b}$$

$$\delta(0 \cdot T_S + 1 \cdot T_s) = \delta(1 \cdot T_s) = 0 \tag{3.5c}$$

$$\delta(1 \cdot T_S + 1 \cdot T_s) = \delta(2 \cdot T_s) = 0 \tag{3.5d}$$

If and only if the argument to the delta is 0, then the delta evaluates to 1, and in this example, that occurs at $n = 1$ (representing that the student is taking the course a year early). We can control the placement of this delta by setting $k = +1$ so that the argument of the delta equals 0 when $n = -1$. The same concept applies for a delay, but in the case of a delay, k will be negative, so in generally a delayed delta would be expressed as

$$\delta(n \cdot T_s - k \cdot T_s) \tag{3.6}$$

The delay operator is used more commonly in audio processing since it is a rare occasion that we have access to future samples, especially in real-time processing. The delay operators will resurface again in Chapter 7 and onward, as it is a critical element to digital filters and spectral processing. We will utilize the delay operator to make a simple Delay Effect in the following programming example.

3.3.1 Programming example: simple delay effect

In this example, you will create an effect that processes an audio file by combining it with an attenuated and delayed version of itself. There are many real-world instances of direct and delayed signals combining. For example, this effect would be used to model first-order reflections found in room acoustics, where the delay amount depends on the amount of extra distance travelled along the reflected path and the gain depends on the amount of absorption happening at the reflection point.

The first step is to load in an audio file and save it as a variable. This audio file can be downloaded from DigitalAudioTheory.com. The vector x has two columns, one for each stereo channel. We will assign the delay amount, k, as well as the gain factor, b, for the delayed signal. The for-loop will process every sample in x plus the amount of delay (so that the output is not truncated). Next, we will compute the current sample, x_n, and the delayed sample, x_d (note that k is negated), and the output, y, is a sum of these.

```
[x,fs]=audioread('snare.wav');
k=10000;
b = 0.25;
N=length(x);

for n=1:N+k
    if n>k && n-k<N
        xd=x(n-k,:);
    else
        xd=[0 0];
    end
    if n<N
        xn = x(n,:);
    else
        xn=[0 0];
    end

    y(n,:)=xn+b*xd;
end

soundsc(y,fs)
```

3.4 Sampling a continuous signal

A delta train can be constructed by summing together many time-shifted individual deltas. Each of these deltas is separated by the sampling period, T_S and is given by

$$d(t)= \sum_{k=-\infty}^{\infty} \delta(t-k \cdot T_s) \qquad (3.7)$$

The index variable, k, is an integer, which forces values of the delta train to be spaced apart by T_S, since the delta function will always evaluate to 0 **unless $t = kT_S$**. For example, when $t = -3T_S$,

$$d(t =-3 \cdot T_S)= \sum_{k=-\infty}^{\infty} \delta(-3 \cdot T_S -k \cdot T_S)= \begin{cases} 1, & k=-3 \\ 0, & otherwise \end{cases} \qquad (3.8)$$

$d(-3T_S) = 1$ if and only if $k = -3$. Importantly, $d(t) = 1$ at **integer multiples of T_S and 0 everywhere else**. Therefore, to sample a continuous function with spacing of T_S (a sample rate of f_S), we will multiply it by $d(t)$.

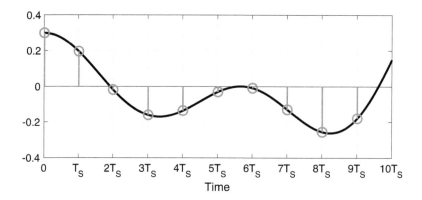

Figure 3.5

A continuous function (black), when multiplied by a delta train produces a "sampled" version of the signal (gray), with a sampling period of T_S.

We can envision the sampling process in the time-domain as a capture of the signal value at equal intervals of T_S. Consider the signal $x_c(t)$ in Figure 3.5, which is sampled by a delta train. The delta train, d, equals 1 at intervals of T_S, so when it is multiplied by a continuous function, then a discrete sequence of samples is generated.

In mathematical form, we can consider a continuous function $x_c(t)$:

$$x_c(t) = \sin(2\pi 2t) \tag{3.9}$$

If we let $T_S = 0.1$ s, then the sampled version, $x_s(t)$, is:

$$x_s(t) = \underbrace{\sin(2\pi 2t)}_{\text{continuous function}} \cdot \underbrace{\sum_{k=-\infty}^{\infty} \delta\left(\overset{n\cdot T_S}{\hat{t}} - k\cdot T_{\bar{s}|}\right)}_{\text{delta train}} \tag{3.10}$$

Recall that the delta function is zero everywhere *except* at $t - kT_S = 0$, therefore, we only capture a value of $x_c(t)$ at these moments. In Table 3.1 the values of $x_s(t)$ at these discrete time points is computed. These values are shown in Figure 3.6 as a digital sequence.

Since $x_s(t)$ is only defined at multiples of T_S (and is zero elsewhere), it can equivalently be written as

$$x_s(t) = x_s(n \cdot T_s) \tag{3.11}$$

This can be simplified even more; $x_s(t)$ is a continuous signal, but it only holds values at discrete time points, so we can equivalently drop T_S from the equations, and define a new operator, the square bracket [...], to indicate a digital sequence, such that $[n]$ is the same as (nT_S).

Table 3.1 For a $T_S = 0.1$ s, the Continuous Function $x_c(t)$ is Sampled, Giving $x_s(t)$. For Each Value of n (column 1), the Delta Train Evaluates to One When Its Argument Is Zero (column 2), Which Occurs at One Specific Time Point (column 3), Resulting in a Value for $x_s(t)$ (column 4)

When $n =$	$d(t) = 1$	when $t =$	$x_s(t)$
0	$t - 0 \cdot T_S = 0$	$t = 0$	$\sin(2\pi \cdot 2 \cdot 0) = 0$
1	$t - 1 \cdot T_S = 0$	$t = T_s$	$\sin(2\pi \cdot 2 \cdot 0.1) = 0.95$
2	$t - 2 \cdot T_S = 0$	$t = 2T_s$	$\sin(2\pi \cdot 2 \cdot 0.2) = 0.59$
3	$t - 3 \cdot T_S = 0$	$t = 3T_s$	$\sin(2\pi \cdot 2 \cdot 0.3) = -0.59$
4	$t - 4 \cdot T_S = 0$	$t = 4T_s$	$\sin(2\pi \cdot 2 \cdot 0.4) = -0.95$
5	$t - 5 \cdot T_S = 0$	$t = 5T_s$	$\sin(2\pi \cdot 2 \cdot 0.5) = 0$

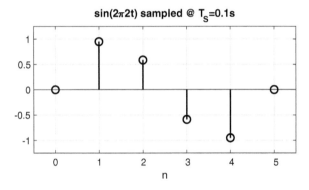

Figure 3.6

The sampled representation of $x_c(t)$ with $T_S = 0.1$s, using values from Table 3.1.

$$x[n] \overset{\text{def}}{=} x_s(n \cdot T_s) = x_c(t) \cdot \sum_{k=-\infty}^{\infty} \delta(n \cdot T_S - k \cdot T_S) \tag{3.12}$$

This raises one important question – what is an appropriate value for T_S to ensure a proper representation of $x_c(t)$? As we saw in Section 3.2, the answer is quite simply this: the minimum number of samples required to capture a complete cycle of a waveform is >2. In other words,

$$T_S < \frac{1}{2} T_{min} \tag{3.13a}$$

or equivalently,

$$f_S > 2 f_{max} \tag{3.13b}$$

where T_{min} is the period of the highest frequency, f_{max}, in the continuous signal $x_c(t)$. This relationship is known as Nyquist sampling theorem. Since the human ear is capable of hearing up to 20 kHz, then the minimum sampling rate to reproduce all audible frequencies must be greater than twice this, or at least 40 kHz. A typical audio sampling rate is 44.1 kHz, putting the Nyquist frequency at 22.05 kHz – this gives ample room for an LPF to maintain 20 kHz in the pass-band, while sufficiently rejecting all frequencies above 22.05 kHz. The reason for such an LPF is discussed in Chapter 4.

3.4.1 Example: analog to digital conversion

Problem: Write the equation that describes the digital signal $x_s[n]$ for the following continuous-time signal sampled at 44,100 Hz:

$$x_c(t) = \sin(2\pi 441t) + \cos(2\pi 882t)$$

Solution: Recall that we replace t with nT_S, and that $T_S = 1/f_S$. Therefore, we can write

$$x_s[n] = \sin\left(\frac{2\pi 441n}{44{,}100}\right) + \cos\left(\frac{2\pi 882n}{44{,}100}\right)$$

$$= \sin(0.02\pi n) + \cos(0.04\pi n)$$

Double Check: The first 4 values ($n=0...3$) of $x_s[n]$ are:

$$x_s[0,1,2\ 3] = \sin(0.02 \cdot \pi \cdot [0,1,2,3]) + \cos(0.04 \cdot \pi \cdot [0\ 1,2,3])$$

$$x_s = [1.000,\ 1.055,\ 1.094,\ 1.117]$$

The values of $x_c(t)$ for $t=0$, T_S, $2T_S$, $3T_S$ are:

$$x_c([0,T_S,2T_S,3T_S]) = \sin(2\pi 441 \cdot [0,T_S,2T_S,3T_S])$$
$$+\cos(2\pi 882 \cdot [0,T_S,2T_S,3T_S])$$

$$x_c = [1.000,\ 1.055,\ 1.094,\ 1.117]$$

3.5 Jitter

The sampler is governed by a high precision clock that acts as a metronome. At precise time intervals (T_S), the clock is responsible for indicating the exact moment that the signal level must be captured. For the sampled signal to be perfectly reconstructed, the time interval between each successive sample *must be* identical. Deviations from perfectly isochronic sampling are known as *jitter*. If a sample is captured at an imprecise time, then the signal will become distorted, depending on the nature of the jitter. Music/audio engineers may already be familiar with temporal non-linearities from the analog domain. For example, both tape as well as turntable playback systems, can suffer from time variations, often described as wow and flutter, often heard as a frequency wobble. However, digital jitter has a different impact, adding noise and distortion to the sampled signal. And if jitter is present on the ADC, then the signal becomes irrecoverably distorted.

The impact of jitter depends on the rate of change of the signal – in other words, large-amplitude and high-frequency signals are the most prone to jitter distortions. There are many possible sources of jitter, in addition to the converter clock – jitter can arise from interconnections between digital devices, for example, an external clock sync. And even if the digital signal was perfectly sampled, if the DAC has jitter, then increased noise and distortion will be introduced on playback. Additionally, if each of these systems has jitter, the total jitter noise within the signal will accumulate. Jitter can be characterized by three major manifestations. If the jitter is randomly distributed around the true sampling interval, then the jitter is random and results in an increase in the noise floor. If the jitter is periodic, it results in the addition of a constant tone (and harmonic series) to the signal. Finally, FM jitter results in *sideband* frequencies, or extra frequencies that are added to the signal that moves with the frequencies originally present within the signal.

Jitter is characterized by the standard deviation of the timing intervals between successive time samples (in the case of random jitter) – this can identically be thought of as the RMS of the timing error. It is a reasonable question to ask, "what is a tolerable amount of jitter for a digital audio system?" We can address this question by considering the worst-case scenario, or a very high frequency, very high amplitude signal – let's use the Nyquist frequency at 0 dB-FS, which for a 16-bit digital system, gives a dynamic range of 96 dB. We want the jitter noise floor to be *no greater* than the dynamic range of our converter or digital system. For the jitter to have an SNR greater than 96 dB, the amount of jitter with a standard deviation of t_j for a sinusoid with frequency, f, is [1]:

$$t_j = \frac{10^{-\frac{SNR}{20}}}{2\pi f} = \frac{10^{-\frac{96}{20}}}{2\pi \cdot 22{,}050} = 0.1 \ ns \qquad (3.14)$$

While jitter is impossible to avoid completely, this provides some guidelines as to what amount of jitter is tolerable in order to guarantee that it does not contribute in any significant way to the overall noise floor, even in the worst-case scenario. You may be wondering what jitter sounds like, and this will be explored in the following programming example.

3.5.1 Programming example: listening to jitter

In this example, you will simulate jitter and listen to its impact. While this example uses a single 1 kHz sinusoid, you could also experiment with loading in an audio file and increasing its sampling rate by a factor of *src* (sample rate conversion) using the function **resample**(). Here, we will make the standard of deviation of the jitter, j_{RMS}, a variable to allow for auditioning of several different amounts. The process of sampling will be simulated by designing a 1 kHz sinusoid at a sample rate of 4.41 MHz (to simulate the analog signal), then keeping every 100th sample, resulting in a sample rate 4.41 MHz / 100 = 44.1 kHz. Except, instead of taking exactly every *src* = 100th sample, a random offset with a standard deviation of j_{RMS} will be applied, resulting in "sampling" intervals of approximately every 100th sample, sometimes more and sometimes less. The jitter offset is to be randomly generated with the function **randn**(), which follows a Gaussian pdf. Can you hear jitter at these varying amounts? 10 ns; 100 ns; 1,000 ns.

```
% sample rate of 44.1 kHz
Fs=44100;

% sample rate conversion factor
src = 100;

% create the 1 kHz sinusoid
x=sin(2*pi*1000*(0:Fs*src)/Fs/src);

% set the std. dev. of the jitter in nano-sec
j_ns = 1000;

% convert from ns to samples
j_rms = j_ns*1e-9*Fs*src;

% sample index for 44.1 kHz
cnt=1;

% iterate over the entire digital signal, in skips of src
for n=1:src:length(x)

    % generate a random number
    jit = randn();

    % 1) scale the jitter amount by j_rms
    % 2) convert to an integer
    % 3) make sure index is in-bounds
    idx = max(1, round(n+j_rms*jit));
    idx = min(idx, length(x));

    % capture the value of x at jittered intervals
    y(cnt) = x(idx);

    % increment the sample index (44.1 kHz)
    cnt=cnt+1;
end

sound(y,Fs)
```

3.6 Challenges

1. State the Nyquist-Shannon sampling theorem mathematically
2. A microphone signal contains frequencies up to f_{max} = 18,000 Hz. What is the minimum sampling rate that you could select?
3. Given a signal $x_c(t) = \cos(2\pi 100t)$ and a sampling rate of f_S = 500 Hz.
   a. Calculate the first five terms of $x[n]$ (Hint: substitute nT_S for t).
   b. Determine a closed form expression for $x[n]$
4. Plot x in "Frequency Space" (Hint: x is real, so how many phasors does it comprise?)
5. Construct a T = 1 s digital signal, $x[n]$, in MATLAB® or Octave with a frequency of 1,000 Hz and a sampling rate of f_S = 44,100 Hz. (Hint: $t = nT_S$, where n is a vector of values from 1 to Tf_S). Plot the first 0.1 s of x.
6. A continuous signal is given by $x = R \cdot \cos 2\pi ft$. Prove that this is equal to $\frac{1}{2}\left(R \cdot e^{j2\pi ft} + R \cdot e^{-j2\pi ft}\right)$.
7. An ADC clock exhibits random jitter with an RMS of 0.25 ns. What is the SNR for this amount of clock jitter?
8. Create both a cosine and sine signal with a frequency of 500 Hz and a sampling rate of f_S = 1,000 Hz with a duration of 0.1 s. Plot both of these signals on the same graph. Explain the reason for the difference between these.

Bibliography

[1] Brannon, B., "Sampled Systems and the Effects of Clock Phase Noise and Jitter," *Analog Devices Inc.* Application Note AN-756, 2004.

4

Aliasing and reconstruction

If you completed Challenge #8 from Chapter 3, you saw that for a cosine function with a frequency of exactly $f_S/2$, it was representable **if and only if** its starting phase was precisely 0 rad. When plotting at a different starting phase, as we did with the sine function that differs from a cosine by $\pi/2$ rad, it was not recoverable since only the zero-crossings were captured by the sampling process! Strictly speaking, this sinusoid violated the Nyquist Theorem since its frequency was not less than f_N (it was exactly f_N). This inability to appropriately capture a signal whose frequency is $\geq f_N$ results in a distortion known as *aliasing*. It wouldn't be so egregious if the sampling process somehow ignored frequencies $\geq f_N$, but that is not what happens – instead, those frequencies appear in the digital signal but as lower frequencies! In this chapter, we will explore this phenomenon, as well as a filter for preventing aliasing, known as *anti-aliasing filter*.

Interestingly, while the ADC capture of a signal with frequencies $\geq f_N$ results in them appearing below f_N in the digital signal, during the DAC reconstruction of a digital signal, frequencies $< f_N$ can be reproduced as frequencies $> f_N$ – how is that for a brain teaser! This has to do with the fact that samples arrive at the DAC as a series of impulses with zeros in between them. But this is not representative of continuous, analog audio; therefore, during reconstruction, there is a necessary step of interpolating in between sample values to "fill in" values that are otherwise not defined. There is a type of filter, known as an *interpolation filter*, that accomplishes this, and interestingly the interpolation filter is identical to the anti-aliasing filter. This chapter includes a programming example in which you will sample and reconstruct an audio signal and concludes with a project in which you will listen to the audible effects of aliasing.

4.1 Under-sampling

We saw in Chapter 3 that if a phasor or sinusoid undergoes a phase rotation of more than π rad during the sampling period, T_S, then the interpretation of its frequency is ambiguous and unrecoverable. (Recall: for a phasor with a frequency $\geq f_N$, it appeared on the complex plane as a slower frequency with negative frequency, or opposite phase). For a signal comprising many frequencies (e.g., speech and music), it can be unambiguously represented if (and only if) its maximum frequency, f_{max}, is strictly $< f_S/2$. Let's look at an example of aliasing. Consider two sinusoids, x_1 and x_2, with frequencies of 2 and 8 Hz, respectively:

$$x_1(t) = \sin(2 \cdot \pi \cdot 2 \cdot t) \tag{4.1a}$$

$$x_2(t) = \sin(2 \cdot \pi \cdot 8 \cdot t) \tag{4.1b}$$

If we set the sample rate to $f_S = 10$ Hz ($T_S = 0.1$ s), then based on Nyquist sampling theorem, we already know $f_{max} < 5$ Hz. Therefore, we expect something odd to happen for x_2. Let's examine both of these in detail in Table 4.1.

As we saw in Chapter 3, we construct our digital sequence by replacing t with nT_S; iterating over the first few values of n, we actually get one complete cycle for $x_1[n]$ and $x_2[n]$. If we plot the digital sequences $x_1[n]$ and $x_2[n]$ it can be seen that both appear as 2 Hz sinusoids! In fact, there is a symmetry between $x_1[n]$ and $x_2[n]$ – note that they mirror each other, when x_1 increases, x_2 decreases by the same rate. Let's explore this with the following programming example.

4.1.1 Programming example: aliasing

In this example, we will sample two sinusoids, x_1 and x_2 with frequencies of 2 Hz and 10 Hz, respectively, using a sampling rate of 10 Hz.

Table 4.1 Examination of Two Signals, x_1 at 2 Hz and x_2 at 8 Hz, Both Sampled at $f_S = 10$ Hz. It Is Noted That x_2 Exceeds the Nyquist Rate of $10/2 = 5$ Hz.

When $n =$	$x_1[n] =$	$x_2[n] =$
0	$\sin\left(2\pi \cdot \underset{f}{2} \cdot \underset{n}{0} \cdot \underset{T_S}{0.1}\right) = 0.00$	$\sin\left(2\pi \cdot \underset{f}{8} \cdot \underset{n}{0} \cdot \underset{T_S}{0.1}\right) = 0.00$
1	$\sin(2\pi \cdot 2 \cdot 1 \cdot 0.1) = 0.95$	$\sin(2\pi \cdot 8 \cdot 1 \cdot 0.1) = 0.95$
2	$\sin(2\pi \cdot 2 \cdot 2 \cdot 0.1) = 0.59$	$\sin(2\pi \cdot 8 \cdot 2 \cdot 0.1) = 0.59$
3	$\sin(2\pi \cdot 2 \cdot 3 \cdot 0.1) = -0.59$	$\sin(2\pi \cdot 8 \cdot 3 \cdot 0.1) = -0.59$
4	$\sin(2\pi \cdot 2 \cdot 4 \cdot 0.1) = -0.95$	$\sin(2\pi \cdot 8 \cdot 4 \cdot 0.1) = -0.95$
5	$\sin(2\pi \cdot 2 \cdot 5 \cdot 0.1) = 0.00$	$\sin(2\pi \cdot 8 \cdot 5 \cdot 0.1) = 0.00$

4. Aliasing and reconstruction

```
Fs=10;          % sample rate = 10 Hz
T = 1;          % duration in s
n=[0:T*Fs];     % let's look at the first second

f1=2;           % x1 frequency 2Hz
f2=8;           % x2 frequency 8Hz

x1=sin(2*pi*f1*n/Fs);
x2=sin(2*pi*f2*n/Fs);

plot(n/Fs,x2);
grid on;
hold on;
plot(n/Fs,x1);

% Let's also look at x2 sampled appropriately
Fs_hi=1000; % sample rate = 1000 Hz
n_hi=[0:T*Fs_hi];
xc=sin(2*pi*f2*n_hi/Fs_hi);
plot(n_hi/Fs_hi, xc, '--')
```

In Figure 4.1, you can see two digital signals, $x_1[n]$ (black) and $x_2[n]$ (gray) in thin lines with dots indicating samples at intervals of $T_S = 0.1$ s, in addition to one continuous signal, $x_2(t)$. Since the frequency of $x_1(t)$ is $<f_N$ then it is accurately captured in the sampling process – you can count that there are two complete cycles in 1 s, or 2 Hz. However, the sampling rate is not high enough to capture the amount of

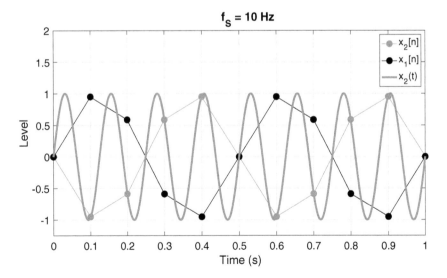

Figure 4.1

For a sampling rate of 10 Hz, the Nyquist frequency is 5 Hz. Therefore a 2 Hz signal, $x_1(t)$, is appropriately sampled (black) while an 8 Hz signal, $x_2(t)$ shown in gray (thick), is inappropriately sampled and appears as an aliased frequency (gray).

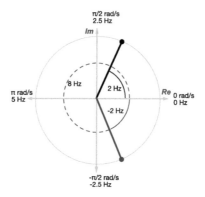

Figure 4.2

Here we see a unit circle in Frequency Space for a sampling rate of $f_S = 10$ Hz. A 2 Hz phasor (black) and an 8 Hz phasor (gray) are shown. While the 2 Hz phasor is appropriately sampled, evidenced by that fact that it appears in the upper half of the unit circle, the 8 Hz signal **aliases** as a –2 Hz phasor (in other words, the phase and the imaginary parts are opposite of the 2 Hz phasor).

change occurring in $x_2(t)$, therefore $x_2[n]$ does not appear as an 8 Hz sinusoid, but rather a 2 Hz one.

The 2 Hz exhibited by $x_2[n]$ is said to be an alias of $x_2(t)$ since it does not truly represent the original signal. Another interesting thing about the aliased $x_2[n]$ is that it is phase shifted by π rad compared to $x_1[n]$. This is easily explained by plotting these two sinusoids in frequency space on the complex plane. But instead of examining two real signals, $\sin(2\pi 2nT_S)$ and $\sin(2\pi 8nT_S)$, both of which produce two phasors each on the complex plane, for the purposes of demonstration, let's consider two complex digital signals with the same frequencies: $e^{j2\pi 2nT_S}$ and $e^{j2\pi 8nT_S}$. Consider Figure 4.2, first examining the axis labels, it is clearly in frequency space since the axes are labeled with frequencies corresponding to the sampling rate, whereby π rad/s $\rightarrow f_N = 5$ Hz. Here, you can see why an 8 Hz signal, sampled at 10 Hz, aliases as a 2 Hz signal with opposite phase.

4.2 Predicting the alias frequency

A tautological yet relevant fact about a circle is that you can rotate around it as many times as you please. In the previous example, we saw that a phasor aliased as soon as it entered into the bottom half of the unit circle. But we can also anticipate aliasing for a phasor that rotates all the way around the unit circle and ends up back in the upper half, which is to say that its frequency is greater than f_S but less than $3f_S/2$ (or greater than $2f_S$ but less than $5f_S/2$, or any other infinite possibilities as well). Consider Figure 4.3 (a), the black phasor we have to assume is 0 Hz, but we saw previously that a frequency that is exactly f_S Hz can also alias as a 0 Hz

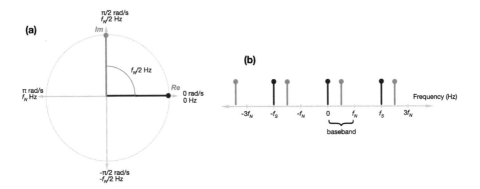

Figure 4.3

A frequency that is a multiple of f_S Hz (black) when sampled will alias as 0 Hz, while frequencies at $f_N/2 \pm f_S$ Hz (gray) will alias as $f_N/2$ Hz. (a) shows this phenomenon on the unit circle while (b) shows this on a frequency line. The aliased frequency will always appear within baseband, or 0 to fN Hz.

signal (but then so could a frequency that is $2f_S$ Hz, or $3 f_S$ Hz, and so on). If we were to "unravel" the unit circle and straighten it, it would appear as a line segment from $-f_N$ to $+f_N$, which represents one complete circumnavigation around the unit circle. However, as we saw, an aliasing phasor may make more than one complete trip around the unit circle, so we can expand the frequency line to account for this, as shown in Figure 4.3(b). Each vertical black line segment represents a possible frequency for the black phasor in Figure 4.3 (a) at 0 Hz, which is to say that any of these frequencies will alias as a 0 Hz signal. The same phenomenon occurs with a phasor of a different frequency, this time $f_N/2$, which is represented in gray in Figure 4.3(a) and (b).

Indicated in Figure 4.3 (b) is *baseband* – or the set frequencies that can be represented by a given sampling rate. This frequency range by definition includes 0 Hz and includes the frequencies up to the Nyquist rate, f_N. Recall that a real signal results in a pair of complex phasors, one in baseband and a conjugate that appears in the lower half of the unit circle. But this region could be interpreted as either $-f_N$ Hz to 0 Hz or equally as f_N Hz to f_S Hz. And, in fact, these frequencies are digitally equivalent. This is simply the reality of dealing with a digital signal. The long and short of it is this: all frequencies within a sampled signal that exist outside of baseband will appear digitally within baseband as an aliased frequency.

While the frequencies that produce aliases in Figure 4.3 are f_S Hz apart, we already saw that frequencies between f_N and f_S also result in aliases. In fact, a signal with a frequency of 22,051 Hz sampled at 44.1 kHz will result in an alias at 22,049 Hz, only a 2 Hz difference. That is, the alias will appear Δf Hz below f_N for a sampled sinusoid whose frequency is Δf Hz above f_N. This can be visualized if we consider an audio source with a spectrum with most of the energy that rolls-off from low to high frequencies, as shown in Figure 4.4. This spectrum is visible as a triangle from 0 Hz to f_N Hz. Any other frequency regions that exhibit this shape with

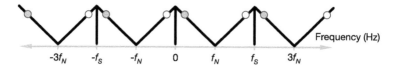

Figure 4.4

A pink magnitude spectrum is shown from 0 Hz to f_N Hz. This spectrum is reversed from 0 to $-f_N$ Hz, since real signals also include conjugates (on the lower half of the unit circle) that match in magnitude but with opposite phase. This magnitude spectrum is repeated every f_S, corresponding with rotations around the unit circle. All frequencies indicated with a dot on the spectrum will appear as $f_N/4$ Hz digitally. Filled dots indicate the alias will be in phase and unfilled dots indicate it will be out of phase.

the same orientation (for example, from f_S to f_S+f_N Hz) will alias as a frequency in baseband. However, other frequency regions exhibit a magnitude spectrum that is a right triangle, but flipped horizontally (for example, from f_N to f_S). These frequencies will also alias within baseband, but with opposite phase. In Figure 4.4, all of the frequencies indicated with a filled dot will alias as $f_N/4$, while all of the frequencies indicated with an unfilled dot will alias as the conjugate of $f_N/4$ (in other words, with opposite phase, just like we saw in the programming example 4.1.1.).

There is an equation that can be used to determine what the aliasing frequency, f_a, will be given the underlying frequency, f, and the sampling rate, f_S, given by:

$$f_a(band) = |f - band \cdot f_S|$$ (4.2)

Where *band* is the bandwidth of frequencies that f falls in, as shown in Figure 4.5.

The following provides a step by step guide to determining what the alias frequency, f_a, will be.

Step 1 – Determine the *band* of f. Baseband is defined as the frequencies up to f_N. Each subsequent band spans f_S Hz with increasing band number (*band*), as shown in Figure 4.5.

Step 2 – Solve for f_a. We can work out some simplified equations for various bands. Note that at baseband, aliasing does not occur (Table 4.2).

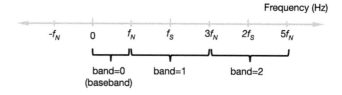

Figure 4.5

Any frequency appearing in a band other than baseband (*band* = 0) will result in an alias frequency within baseband. Each subsequent band straddles integer multiples of f_S.

4. Aliasing and reconstruction

Table 4.2 Aliasing Frequency for Different Frequency Ranges

band	f	f_a
0	$f < f_N$	$f_a = f$
1	$f_N \le f < 3f_N$	$f_a = \|f - f_s\|$
2	$3f_N \le f < 5f_N$	$f_a = \|f - 2f_s\|$

4.2.1 Example: calculating the alias frequencies

Problem: You are digitally archiving some tape with an AC bias of 50 kHz. This signal is inaudible in the analog domain – what is the alias frequency of the AC bias if sampled at $f_S = 44.1$ kHz.

Solution: The first step is to determine which band 50 kHz is in. According to Figure 4.5, it can be determined that $band = 1$. Next, we solve f_a:

$$f_a(50) = |50 - 1 \cdot 44.1| = 5.9 \ kHz$$

Double Check: It is possible to visualize f_a on the unit circle to double-check the calculation. If we rotate a phasor all the way around the unit circle, we end up at 44.1 kHz. It takes another 5.9 kHz to end up at 50 kHz. From Figure 4.6, it is obvious why the alias frequency is 5.9 kHz.

4.2.2 Mirror/foldback frequency

Sometimes the Nyquist is referred to as the *mirror* frequency or *fold-back* frequency. That is because aliasing, when viewed strictly within baseband, appears as a "reflection" or "folding" about both $f_S/2$ Hz as well as 0 Hz. Since this happens at both 0 Hz

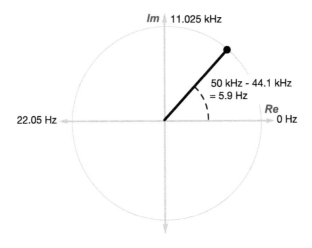

Figure 4.6

A 50 kHz frequency appears as 5.9 kHz when sampled at $f_S = 44.1$ kHz.

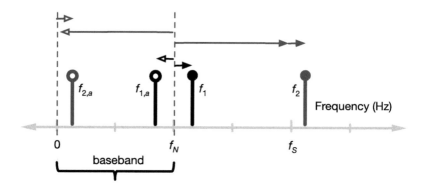

Figure 4.7

A frequency f_1 (black, filled) is above the Nyquist rate, so it aliases at $f_{1,a}$ (black, un-filled), which mirrors f_1 about f_N. Since f_2 is above f_S, its alias, $f_{2,a}$ also mirrors off of 0 Hz.

and Nyquist, aliases are like ping-pong balls bouncing off of these two frequency locations, as shown in Figure 4.7. Fortunately, this ping-pong effect is manifest in the aliasing equation, as we saw in example 4.2.1, in which the alias mirrored off of both 22,050 Hz as well as 0 Hz, much like $f_{2,a}$ in Figure 4.7.

Up until this point, we have been considering complex phasors, that is to say, phasors that have both real and imaginary parts. But now it is worthwhile to take one final look at aliasing with real phasors, since this is what will always be encountered when dealing with audio in the time domain. Recall that a real phasor has its magnitude split between two complex phasors that are conjugates of one another. Consider a real phasor, x_r, with frequency f and magnitude A:

$$x_r[n] = \frac{1}{2}Ae^{j2\pi f n T_S} + \frac{1}{2}Ae^{-j2\pi f n T_S} \tag{4.3}$$

Simply rearranging the variables in the exponent and substituting $1/f_S$ for T_S:

$$x_r[n] = \frac{1}{2}Ae^{j2\pi\left(\frac{f}{f_S}\right)n} + \underbrace{\frac{1}{2}Ae^{-j2\pi\left(\frac{f}{f_S}\right)n}}_{conjugate} \tag{4.4}$$

It can be seen that f/f_S gives a fraction of the unit circle (comprising 2π rad). If properly sampled, this ratio will strictly be <0.5, whereby the positive-phase part (the left term of the right-hand side of Equation 4.4) will always be in the upper half of the unit circle, while the negative-phase part will always be in the lower half of the unit circle. However, when a signal aliases such that the ratio $f/f_S > 0.5$, then the positive-phase part ends up in the bottom half of the unit circle and the negative-phase part in the upper half. This is an alternate explanation for the phase reversal that occurs when an alias mirrors off of f_N. Figure 4.8 (a) shows a 15 kHz

4. Aliasing and reconstruction

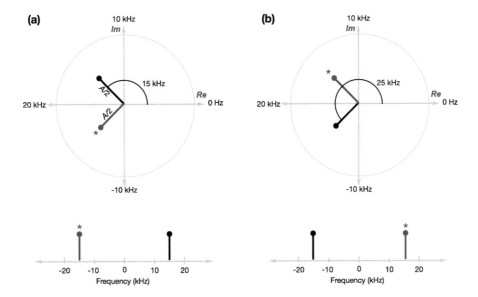

Figure 4.8

(a) A 15 kHz real phasor appropriately sampled at 40 kHz, with magnitude split between a conjugate pair. (b) A 25 kHz real phasor under sampled (also at 40 kHz), with its conjugate appearing in the top half of the unit circle.

real phasor properly sampled at 40 kHz while (b) shows a 25 kHz phasor also sampled at 40 kHz that appears as an alias at 15 kHz. The signals in Figure 4.8 (a) and (b) will be opposite phase of one another even though they are the same frequency.

4.2.3 Example: calculating the alias frequencies (again)

Problem I: Given a continuous signal $x_c(t)$, find $x[n]$ for a sample rate of $f_S = 20$ kHz, where:

$$x_c(t) = \sin(24{,}000\pi t)$$

Solution: To solve for $x[n]$, substitute nT_S for t:

$$x[n] = x_c(nT_S)$$

$$= \sin\left(24{,}000\pi n \frac{1}{20{,}000}\right)$$

$$= \sin(2\pi 0.6 n)$$

Problem II: Given the observed digital signal $x[n] = \cos\left(\dfrac{\pi}{4}n\right)$, determine two possible frequencies, f, for the sampled signal $x_c(t) = \cos(2\pi ft)$ if it was sampled at 1 kHz.

Solution: First we will consider a possible value for f assuming it is sampled appropriately. The arguments of the cosines from both equations can be equated if we substitute nT_S for t and solve for f:

$$2\pi f \overbrace{\left(n\frac{1}{1000} \right)}^{\text{from } x_c(t)} = \overbrace{\frac{\pi}{4}n}^{\text{from } x[n]}$$

where t is indicated under $n\frac{1}{1000}$.

$$f = 125 Hz$$

Alternatively, we could consider that the frequency f is aliasing as 125 Hz. If f is greater than 2π by $\pi/4$ rad/s, then it can be written as follows:

$$2\pi fn\frac{1}{1000} = \left(2\pi + \frac{\pi}{4} \right)n$$

$$f = 1,125 Hz$$

Last, a frequency does not have to surpass f_S to produce an alias; f could be mirrored off of f_N with its negative phase part appearing at $\pi/4$ rad/s.

$$2\pi fn\frac{1}{1000} = \left(2\pi - \frac{\pi}{4} \right)n$$

$$f = 875 Hz$$

4.3 Anti-aliasing filter

It now should be apparent that aliasing is a distortion that is unacceptable for the digital conversion process. In principle this seems simple, we should employ an LPF that allows frequencies below f_N to pass while blocking frequencies above f_N. This type of LPF is known as an *anti-aliasing filter* and is a part of the ADC process, and can be defined according to the following equation:

$$H_{aa}(f) = \begin{cases} 1, -f_N < f < f_N \\ 0, otherwise \end{cases} \tag{4.5}$$

Note, we must include negative frequencies down to $-f_N$ in order to capture the bottom half of the unit circle. However, such an idealized filter, sometimes referred to as a *brick wall filter*, is not possible to realize in practice. The steeper the roll-off, the higher the filter's Q becomes, leading to more complex and expensive designs. Therefore, for the anti-aliasing filter, it must be given an ample transition band in which the filter rolls off to full attenuation. Although we only want frequencies in the passband, we must allow f_N to also encompass the transition band to prevent aliasing of these frequencies. For example, 20 kHz is the upper bound for human hearing, which is where an anti-aliasing filter will begin to attenuate, with 2.05 kHz

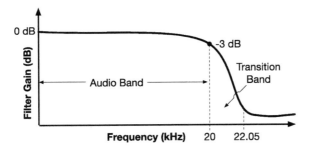

Figure 4.9

Anti-aliasing filter for a 44.1 kHz sample rate. The passband of this LPF extends up to 20 kHz, with a 2.05 kHz transition band.

of bandwidth for the transition band. This puts f_N at 22.05 kHz, which gives us the common audio sampling rate of 44.1 kHz (see Figure 4.9). The anti-aliasing filter must attenuate frequencies greater than f_N prior to the digital conversion process, which is to say the anti-aliasing filter is an analog filter.

4.4 Reconstruction

It may already be apparent that a digital signal is not appropriate for reconstruction. As opposed to a continuous waveform, a digital signal is a train of samples interspersed with zeros. In addition to this, the frequency spectrum of a digital signal is repetitive, creating energy at high frequencies that simply did not exist in the original analog signal. Recall from the introduction to Chapter 4, it was stated that during the DAC reconstruction that frequencies below f_N can be reproduced in frequencies $>f_N$. This claim can be substantiated by considering Figures 4.3 (b) and 4.4. Due to the fact that we can circumnavigate the unit circle as many times as we please, a phasor with a frequency $<f_N$ will appear again and again at higher and higher frequencies with each revolution. This results in a strange situation where the spectrum within baseband is replicated at integer multiples of f_S. These repeated spectra (apart from baseband) are known as *images*.

We've so far identified two problems with the digital signal that make it unsuitable for analog representation: (1) the signal in the time domain is an impulse train, and (2) in the frequency domain, there are images outside of baseband. These two problems are in fact related, so if we solve one, we actually solve both. To address the first problem, a continuous signal could be generated if there were values interpolated in-between samples. Use of interpolation would "fill in" values where none exist. And for the second problem, an LPF to remove images outside baseband would be warranted. The ultimate goal is to convert a digital signal $x[n]$ into a continuous signal, x_r, using either an interpolator, h_r, or a reconstruction LPF, H_r. Let's start by defining H_r:

$$H_r(f) = \begin{cases} 1, -f_N < f < f_N \\ 0, otherwise \end{cases} \qquad (4.6)$$

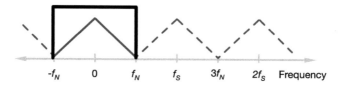

$-f_N$ 0 f_N f_s $3f_N$ $2f_s$ Frequency

Figure 4.10

Digital images (dashed) outside baseband (solid gray) are removed by an ideal analog reconstruction filter, H_r (black).

Not coincidentally, H_r is defined exactly the same as the anti-aliasing filter H_{aa} – they both remove frequencies $\geq f_N$ (and $\leq -f_N$). Figure 4.10 shows a pinkish spectrum in baseband, with images at higher frequencies that are removed with a reconstruction filter.

4.4.1 Deriving the interpolator

If you have no experience with Fourier transforms, then section 4.4.1 can be skipped and you can continue with 4.4.2. Here, we will derive the sample interpolator from the reconstruction filter H_r. If we want to convert H_r to the time domain, then we can use the inverse Fourier transform. We will use the limits $\pm \omega_N$ (where $\omega_N = 2\pi f_N$) on the integral since the LPF is only defined in baseband. The scalar "1" represents the magnitude (equivalent to 0 dB) of the LPF in baseband:

$$h_r(t) = \mathscr{F}^{-1}(H_r(\omega)) \triangleq \frac{1}{2\pi} \int_{-\infty}^{\infty} H_r(\omega) \cdot e^{j\omega t} d\omega \tag{4.7a}$$

$$= \frac{1}{2\pi} \int_{-\omega_N}^{\omega_N} 1 \cdot e^{j\omega t} d\omega \tag{4.7b}$$

$$= \frac{1}{2\pi} \cdot \frac{1}{jt} \cdot \left(e^{j\omega t} \right) \Big|_{-\omega_N}^{\omega_N} \tag{4.7c}$$

$$= \frac{1}{2\pi} \cdot \frac{1}{jt} \cdot \left(e^{j 2\pi f_N t} - e^{-j 2\pi f_N t} \right) \tag{4.7d}$$

Looking at the last equation, if we combine the two complex phasors in the parentheses with 1/2 and 1/j we get Euler's definition of a sine:

$$= \frac{1}{\pi t} \sin(2\pi f_N t) \tag{4.8}$$

     4. Aliasing and reconstruction

Then by multiplying and dividing by f_S (or $2f_N$) we can substitute in a sinc function, which is defined as $\text{sinc}(t) = \sin(\pi t)/\pi t$, giving

$$h_r(t) = f_s \cdot \text{sinc}(tf_s) \tag{4.9}$$

4.4.2 Ideal interpolation

The function that will interpolate between sample values, is a sinc with an argument of tf_S. Let's look at how the sinc evaluates at different points in time around a sample. For the moment, the scale factor f_S can be ignored since the inverse of this scale factor, T_S, can easily be applied at the moment of sampling, so the sinc function can be considered as normalized to one. At time $t=0$, the sinc has an argument of 0, and using l'Hospital's rule it can be seen that the sinc evaluates to 1. At integer multiples of T_S, the argument of the sinc is an integer, meaning that the argument will be an integer multiple of π; therefore, the sinc will always be 0 at these time points. In between time points of T_S, the sinc is sinusoidal, but with a decaying amplitude envelope that is proportional to $1/t$. Such a sinc function, $h_r(t)$, is shown in Figure 4.11.

There is an apparent symmetry between the ADC sampling equation:

$$x[n] = \sum_{k=-\infty}^{\infty} \overbrace{\delta(n \cdot T_S - k \cdot T_s)}^{\text{discrete sampler}} \cdot \overbrace{x_c(t)}^{\substack{\text{continuous} \\ \text{function}}} \tag{4.10}$$

and the DAC reconstruction equation

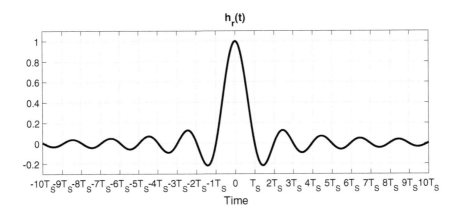

$$h_r(t)$$

Figure 4.11

The reconstruction filter, $h_r(t)$, for a given sample rate is a sinc function with zero-crossings at integer multiples of T_S and a value of 1 and time $t = 0$.

$$x_r(t) = \sum_{n=-\infty}^{\infty} \underbrace{x(n \cdot T_S)}_{\substack{\text{discrete} \\ \text{signal } x[n]}} \cdot \underbrace{\text{sinc}\left(f_S\left(t - n \cdot T_S\right)\right)}_{\substack{\text{continuous} \\ \text{interpolator}}} \qquad (4.11)$$

In the sampling equation, we multiply a continuous signal by a discrete sampler to obtain a discrete output, $x[n]$. With the reconstruction equation, each sample in $x[n]$ is multiplied by a continuous interpolator, and these are all summed together to create the continuous reconstructed output, $x_r(t)$.

Consider the sum that is looping over all n in Equation (4.11). This accumulation happens for every moment in time, t, in other words, **every sample in $x[n]$ contributes to all time points in $x_r(t)$!** That implies that the very last sample in a song contributes to the very first moment in time, although only very minutely. Think of it like this – for every moment in time, we want to add the sinc contribution from every sample in the sequence, but, the further away the sample is from that moment in time, the less impact it will have. The sinc functions provide interpolation between sample locations, where the digital signal is undefined. But right at a value of $t = nT_S$ the value of sinc at <u>that</u> n is 1 and all other sincs are at zero-crossings. In other words, when converting from digital signal to analog signal, if t falls exactly at a sample location, then only that sample value is considered and all others are ignored. But, if t is between samples, then there are contributions from all other samples, weighted by their proximity to t, as shown in Figure 4.12.

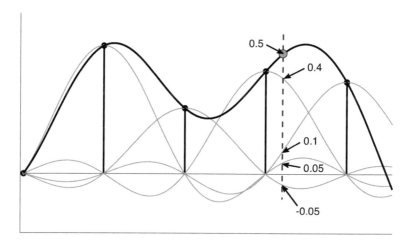

Figure 4.12

Consider the four samples (black dots), each with an associated sinc function (gray), which provide interpolation between sample values. The fully reconstructed signal (black) comprises a sum of each of these sincs. Consider the point marked with a gray dot with a value 0.5. Its value is greater than the value of its two nearest samples; however, the sum of sincs produces the correct value (0.4 + 0.1 + 0.05 − 0.05 = 0.5).

4. Aliasing and reconstruction

4.4.3 Real-time interpolation

In real-time reconstruction, information about future samples simply is not known, therefore sinc interpolation is not possible. In practice, most DACs will perform a *zero-order hold* (ZOH), which is to say that basic "interpolation" is done by simply holding the value of the previous sample until a new sample value arrives at the DAC. This process results in the signal appearing as a stair-step, as shown in Figure 4.13 (a), in which a sample value is held for an entire sample period, T_S. Clearly a ZOH is not representative of the underlying signal, and this is reflected in the spectrum as well. It can be seen that a ZOH results in a roll-off in the spectrum that attenuates frequencies $<f_N$, but also allows some digital images to pass, as shown in Figure 4.13 (b).

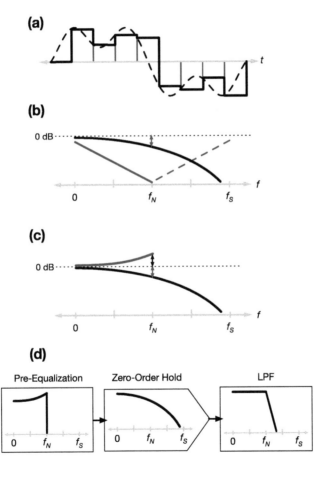

Figure 4.13

(a) A zero-order hold maintains the value of a sample until the next arrives, resulting in a staircase shape. (b) The spectrum of the ZOH (black) attenuates in baseband and also extends into the digital images (dashed gray). (c) A pre-equalization filter (gray) compensates for the loss caused by the ZOH. (d) Signal flow for real-time reconstruction, that ends with an LPF to remove remaining images.

To correct for the ZOH distortions, we must, (1) apply some gain to higher frequencies to correct for the roll-off, and (2) remove frequencies $>f_N$ that are still present in the signal. To address this first problem, we will apply *pre-equalization*, or a filter that counteracts the loss of the ZOH. For example, at a f = $0.8f_N$ the pass band is attenuated by –2.42 dB, therefore, the pre-equalization filter must apply a gain of +2.42 dB at this frequency [1], as shown in Figure 4.13 (c). Importantly, the pre-equalization filter does not need to extend beyond f_N, in fact this would be counterproductive to the reconstruction of an LPF. To address the second problem, some of the work is done for us already, due to the ZOH acting as an LPF. For example, at f_N, there is already attenuation of –8.87 dB [2]. Unlike the steep brick wall filter, this reconstruction LPF can be less aggressive, and therefore more practical to realize. The entire interpolation process for real-time reconstruction is shown in Figure 4.13 (d).

4.5 Challenges

1. A vocal recording contains non-noise frequencies up to 16 kHz, but the ADC is using a sampling rate of f_S = 24,000 Hz.
   a. Indicate what frequencies will alias, and what their aliases will be.
   b. What would be an appropriate cutoff frequency for a low pass anti-aliasing filter?
2. A reconstruction filter has a magnitude spectrum of
$$|H_r(f)| = \begin{cases} 1, -10 \ kHz < f < 10 \ kHz \\ 0, otherwise \end{cases}$$

   Sketch the time-domain interpolator, $h_r(t)$. Label the horizontal axis and ticks.
3. Given a continuous signal $x_c(t) = \cos(2\pi 10,000t)$
   a. Sketch the first 0.5 ms of $x_c(t)$, and label the axes and ticks.
   b. Let f_S = 8,000 Hz. On the same sketch, indicate the location and value of the samples acquired.
   c. Use two of the methods described in the text (aliasing formula, unit circle, or mirroring diagram) to determine the aliased frequency, f_a.
   d. Sketch the frequency location of x_c on a unit circle (in frequency space with f_S = 8,000 Hz). Be sure to label axes. (Hint: x_c is a **real** signal, so think about how many complex phasors should be present).
   e. Why does the alias start in positive polarity and not negative?

4.6 Project – aliasing

Introduction

In this project, you will hear and see the effects of aliasing using swept-frequency sinusoids, also referred to as sine sweeps or chirps. A sine sweep smoothly and monotonically changes frequency, either increasing or decreasing. These come in different flavors that each achieve different goals. A *linear sweep* has constant change in frequency with respect to time; compare this to a *logarithmic sweep*, which has a constant change in chroma with respect to time (in other words, it spans octaves in equal time periods). The logarithmic sweep sounds more natural to the human ear since we perceive pitch based on octave differences rather than absolute frequency differences.

You will observe a sine sweep both appropriately sampled and under sampled using a *spectrogram*. A spectrogram is a visual representation of the spectrum as it changes over time. Imagine the spectrum of a single frequency phasor – on the complex plane it is a vector, and on a horizontal frequency line it is a line segment. We could alternatively represent this as an intensity coded row, almost like rotating the frequency line out of the page and looking down at the top of the line segment. If we repeated this representation and multiple time points, we arrange them vertically into a stack. This allows us to represent the frequency magnitude of a signal as it changes over time, as shown in Figure 4.14.

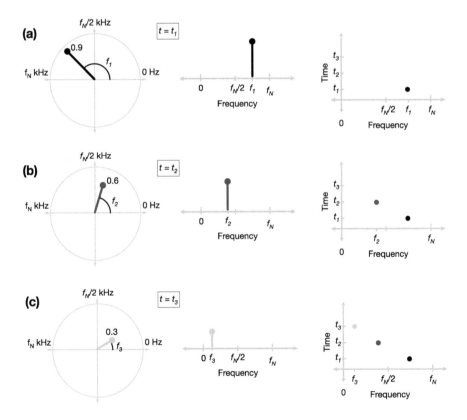

Figure 4.14

A phasor changes frequency (f_1, f_2 and f_3) and magnitude (0.9, 0.6 and 0.3) at three different time points, t_1 (a), t_2 (b) and t_3 (c). The first column shows this phasor on the complex plane, the second column on the frequency line, and the third column shows the spectrogram. The horizontal axis matches the frequency line, while time increases on the vertical axis. Phasor amplitude is indicated by grayscale darkness.

A sine sweep will be generated using the command **chirp**() and will be visualized using the command **spectrogram**(). The chirp function takes as arguments the time vector, start and end frequencies, and the type of sweep – here we will use a linear sweep. The spectrogram function takes arguments for, (1) the signal being analyzed, (2) the amount of time in each analysis window (here, 128 samples at 2,400 Hz is about 53 ms), (3) the amount of overlap of analysis windows (50%), (4) the number of individual frequencies to analyze, and (5) the sample rate.

```
Tmax = 6;      % duration (sec)
Fmax = 600;    % max frequency (Hz)
fs=2400;       % sampling rate (Hz)

t=0:1/fs:Tmax;          % time vector
y=chirp(t,0,Tmax,Fmax); % sine sweep
sound(y,fs)    %listen
spectrogram(y,128,64,512,fs) % see
```

Assignment

    i. Generate a sine sweep from 100 Hz to 1 kHz using a sampling rate of 5 kHz. Listen to it and plot it on the spectrogram.

    ii. Repeat i. with a sample rate of 1.6 kHz. Explain the difference between what is seen and heard for part i.

    iii. What is the minimum sampling rate to accurately capture the sine sweep? Test your hypothesis by repeating part i. with this new f_S.

Bibliography

[1]   Yang, K., "Equalizing techniques flatten DAC frequency response," *Maxim Integrated.* Application Note 3853, 2012.

[2]   Smith, Steven, *The Scientist and Engineer's Guide to Digital Signal Processing,* California Technical Publishing, San Diego, California, 1999, pp. 44–48.

5

Quantization

When an analog signal is sampled, its level must be converted to a digital value. Unlike analog signals, which have continuously varying level, a digital signal can only be represented with finite *resolution*, or the number of discrete representable levels. The resolution is governed by the *bit depth* (or number of bits) of the ADC. For example, if we have a 3-bit encoder, we can possibly represent 8 levels; namely, 000, 001, 010, 011, 100, 101, 110, and 111. With audio signals, we typically use 16-bits (65,536 quantization, or *q*, levels), 20-bits (1,048,576 *q* levels), or 24-bits (16,777,216 *q* levels). Some software packages and DAWs default to *double* type, which simply means two 16-bit words together to create a 32-bit number (4,294,967,296 *q* levels). On the other hand, some music today is processed to recreate the sound of an 8-bit (merely 256 *q* levels) computer synthesizer.

But even if we know the bit-depth of a digital sample, that does not give us any indication of the actual voltage levels that were captured, (voltage being the typical means of transmitting analog audio signals). To **convert** between an analog voltage and a digital level, we need to know a bit more about the converter itself. When we digitize a signal's level, two factors dictate the quantization level, *q*. These are, (1) the range of the analog signal, in other words, the minimum encodable level A^- differenced by the maximum encodable level, A^+, and (2) the bit-depth, N_{bits}, of the converter.

5.1 Quantization resolution

The resolution, q, of an ADC is given by the following equation:

$$q = \frac{A^+ - A^-}{2^{N_{bits}}} \tag{5.1}$$

You can see that there are three terms here: q, A, and N_{bits} – the value of these variables will be selected by the engineer (you!) to fit the needs of the particular application. For example, the quantization level, q, is proportional to the signal's dynamic range; so, if we need to capture a larger dynamic range with a fixed bit-depth, then resolution will suffer. Or conversely, if the resolution needs to be improved, then we either need more bits or a decreased dynamic range. Next, let's consider the bit-depth of the converter itself; we can think of *adding* a bit as *either* improving our resolution (meaning decreasing the quantization level) by a factor of 2x, *or* we can use that extra bit to grow our dynamic range by a factor of 2x.

One problem with quantization is that we cannot guarantee that both the full extent of the dynamic range and zero will be represented by a quantization level. For digital audio, it is necessary that zero is digitally represented, a type of quantization known as a *mid-tread* quantizer. A mid-tread quantizer has a digital value of 0 corresponding to an analog value of 0. The digital value jumps to the next quantization level at an analog value of $\pm q/2$, as shown in Figure 5.1. As a result, the maximum representable value is not A, but rather $A-q$. The quantized value can be predicted according to the quantization function:

$$Q(x) = \left\| \frac{x}{q} \right\| \cdot q \tag{5.2}$$

where $\|...\|$ denotes the function for rounding to the nearest integer, and x is the continuous analog value and q is the quantization step size.

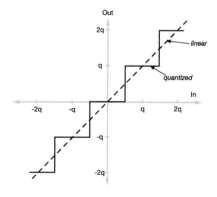

Figure 5.1

Transfer functions of a linear, continuous system (dashed) and a mid-tread quantizer (solid) are shown. Transitions occur at values halfway between steps of q.

5. Quantization

5.2 Audio buffers

When digitized audio is saved to memory from the ADC, it is stored in a *buffer* or a region of physical memory. We can envision the buffer as having rows and columns, where each row represents a sample of digital audio and each column represents the bits from the most significant bit (MSB) at the left to the least significant bit (LSB) at the right. Then each subsequent row represents a different sample, or time-point, as shown in Table 5.1 and Figure 5.2. In real-time digital audio effects processing (e.g., EQ or plug-ins), the DAW will "give" the plug-in a buffer of audio that must be rapidly processed and returned to the DAW before the next buffer is provided.

Table 5.1 For the Quantized Signal Shown in Figure 5.2, the Audio Buffer Will Be Organized in Memory as Shown Here with Adjacent Time Samples Occupying Contiguous Elements in Memory.

	MSB		Buffer	LSB
$t = 0$	1	1	0	1
$t = T_S$	0	0	0	0
$t = 2T_S$	0	0	1	0
$t = 3T_S$	0	0	1	1
$t = 4T_S$	0	0	1	1
$t = 5T_S$	0	0	1	1
$t = 6T_S$	0	0	1	0
$t = 7T_S$	0	0	0	1
$t = 8T_S$	0	0	0	0
$t = 9T_S$	1	1	1	1

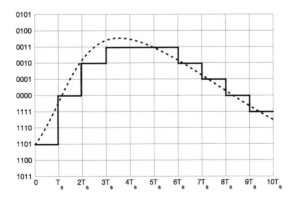

Figure 5.2

A continuous signal (dashed) is quantized with 4 bits (solid) and stored to an audio buffer, as shown in Table 5.1. For each time interval, in multiples of T_S, the quantizer selects the digital level that is closest to the signal level.

You will likely note the abrupt change from the 0 digital value (0000) to the −1 digital value (1111) – this is because most digital audio sequences are stored in signed two's complement form. Under two's complement, the MSB is the 'sign' bit that indicates a positive (0) or negative (1) value; positive values then follow a normal binary sequence, while negative values are stored as signed two's complement of their absolute value. The negative values can be computed by starting with the positive value, inverting all the ones and zeros, then adding one. For example, to represent −2 in two's complement with 4 bits, first, start with the positive value (0010), then invert all the digits (1101) and finally add one (1110). The advantage of using two's complement system is that the basic math operations (addition, subtraction, multiplication) are the same as for unsigned representation.

5.2.1 Programming example: signal quantization

In this programming example, a very high bit-depth signal, which will be treated as essentially continuous, will be quantized using a custom quantization function, that will be called **quantBits**(). In this function, a float value (between −1 and +1) will first be multiplied by the number of quantization levels. Next, that scaled value will be rounded to the nearest integer, and finally, divided by the number of quantization levels. This function effectively quantizes the signal to a bit-depth of N bits.

```
function Q = quantBits(input, N, A)
% Q = quantBits(input, N, A)
% This function quantizes the input according to the bit depth N
for a % signal with dynamic range A

q = (A-(-A))/2^N;
Q = round(input/q)*q;
```

Next, we will use this function to quantize a signal (x) to 8 bits ($x8$). This example signal will be a 4-second cosine with a frequency of 0.25 Hz and a sample rate of 1,000 Hz. By default, this signal will have a bit-depth of 32 bits. A plot will allow comparison of the quantized signal to the original. Additionally, the theoretical quantization step size (q) will be computed, which can be compared to the observed quantization step size from the plot (Figures 5.1 and 5.3).

```
fs=1000;    % sample rate (Hz)
T=4;        % duration (s)
n=0:fs*T;   % sample index vector
f=0.25;     % frequency (Hz)
A=1;        % Max Amplitude
N=8;        % bit-depth

x=0.5*cos(2*pi*f*n/fs); % signal
x8 = quantBits(x,8,A);  % 8-bit quantized version of x

figure; plot(x);
hold on; grid on;
plot(x8);
```

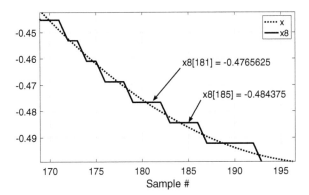

Figure 5.3

The quantized signal (solid) has noticeable steps compared to the original signal (dotted). The level between steps is $q = -0.4765625 + 0.484375 = 0.0078125$, the exact quantization level that we computed for an 8-bit converter over the range $[-1, +1]$.

5.2.2 Example: 3-bit quantization

Problem: A consumer audio line level signal has an RMS voltage of 0.316 V (refer back to Chapter 1.1.2).

  i.   First, determine the dynamic range ($\pm A$) for this level. (Hint: the dynamic range is a <u>peak to peak</u> value).
  ii. Next, determine the quantization levels for a 3-bit converter.
  iii. Finally, quantize the following sequence: $x = \{0.12, -0.02, -0.4, 0.04, 0.10\}$

Solution: To find the amplitude, given the RMS:

$$A = V_{RMS} \cdot \sqrt{2} = 0.316 \cdot \sqrt{2} = 0.447 \text{ V}$$

Next, we can use this range to determine the quantization step size, given a bit-depth of $N_{bits} = 3$ bits:

$$q = \frac{2 \cdot 0.447}{2^3} = 0.11175 \text{ V}$$

This means that the quantization levels in the first column of Table 5.2 correspond to voltages in the second column.

Finally, the quantized sequence can be determined using the following function:

$$Q(x) = \left\| \frac{[0.12, -0.02, -0.4, 0.04, 0.10]}{0.11175} \right\| \cdot 0.11175$$

$$= [0.11175, 0.00000, -0.44700, 0.00000, 0.11175]$$

Table 5.2 For a 3-Bit Converter with a Range of ±0.447 V, the Quantization
Levels in Column 1 Correspond to the Voltage Levels in Column 2.

Digital Level	Quantized Value
011	0.33525 V
010	0.22350 V
001	0.11175 V
000	0.00000 V
111	−0.11175 V
110	−0.22350 V
101	−0.33535 V
100	−0.47700 V

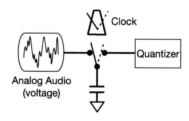

Figure 5.4

A conceptual diagram of a "Sample and Hold" circuit that can be imagined as a
metronome alternating between two throws, one that charges a capacitor with the
level of the analog signal (Left), and one that reads the capacitor for quantizing
(Right). This type of switch is known as a "single pole, double throw" since it has a
single input (the clock in this case) that operates between two gates.

5.3 Sample-and-hold circuit

The sampling process is facilitated with a "sample-and-hold" (S/H) circuit. While
vastly superior designs utilizing op-amps have been devised, a S/H circuit can be
conceptualized as a single-pole, double-throw switch that is governed by the sampling clock (like a very fast metronome). When the clock is high (thought of as
the metronome swinging to the Left), the sample gate is closed, and a capacitor
is charged; then, the clock goes low (the metronome swings to the Right), and the
sampled value is acquired from the capacitor, which is *holding* the sampled value,
as shown in Figure 5.4. The digitized signal then looks like a staircase instead of a
continuously varying level. This is easily visualized by zooming in an audio signal
with your favorite DAW – the discrete levels can be easily seen!

5.4 Quantization error (e_q)

As should be evident by now, the process of quantization introduces error into the
signal, e_q, sometimes also called *granular distortion*. As bit-depth increases the

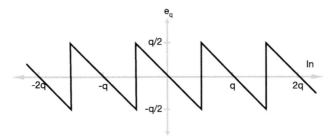

Figure 5.5

The quantization error (also known as granular distortion) falls within the range of $\pm q/2$. This quantization error signal represents the difference between the quantized and analog signals shown in Figure 5.1.

quantization error decreases since the difference between the quantized signal and the original signal gets smaller. Since the value of the analog signal is rounded to the closest quantization level, in the "worst case" scenario, the analog value falls exactly half-way between two quantization levels; therefore, the maximum absolute quantization error is $q/2$, and of course the minimum is 0 in the case that the analog value is exactly on a quantization level. Referring back to Figure 5.1, if we were to difference the quantized value from the analog one, then the error signal (e_q) can be visualized, as shown in Figure 5.5.

5.4.1 Programming example: quantization error

Here, we will plot the quantization error introduced by 8-bit quantization, using the same signals x and $x8$ from the previous programming example. A scaled-down version of the quantized signal itself is also plotted for reference in Figure 5.6.

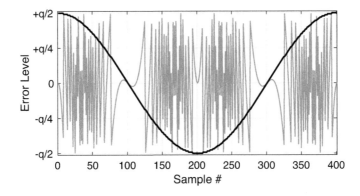

Figure 5.6

The quantization error is the difference between the signal and its quantized version. Note the limits of the vertical axis, the quantization error never goes beyond the range of $[-q/2, +q/2]$.

```
% plot quantization error
eq_x8 = x-x8;
plot(eq_x8)

% plot quantized signal (scaled)
q=2/2^8;
hold on; grid on;
plot(x8*q, 'LineWidth', 3)
yticks([-q/2 -q/4 0 q/4 q/2]);
yticklabels({'-q/2', '-q/4', '0', '+q/4', '+q/2'});
```

Note that in the plot of the quantization error, the limits of the error are within the boundaries of $\pm q/2$, as expected. It can also be seen that e_q is highly correlated to the signal, x. This correlation of e_q to the signal results in a noticeable noise modulation that is sometimes audible, especially for very low-level signals. This problem will be addressed in Chapter 6 using a technique known as dither.

5.4.2 PDF of quantization error

For a refresher on probability density functions, please see Chapter 1.1.7. As we saw previously, an analog value will *sometimes* land *exactly* on a quantization level (when $e_q[n] = 0$). What is the probability of this? Well, let's think of a simple example: a 6-sided die. The probability of throwing any given number is 1 out of 6, or 1/6, where 6 is the total number of possible outcomes. Since the quantization error never goes beyond the bounds of $\pm q/2$, then the total number of possible outcomes for e_q is $q/2 - (-q/2) = q$. Therefore, the probability of any given quantization error level is

$$p(e_q) = \frac{1}{q} \tag{5.3}$$

Since e_q is equally likely to have any value within $\pm q/2$, we say that its PDF is *uniform*, or sometimes called *rectangular* or *boxcar*, and resembles the PDF shown in Figure 5.7.

5.4.3 RMS of quantization error

The RMS level of e_q can be estimated in three steps: (1) squaring the error; (2) taking the mean of the squared error (defined as the integral of the signal times its PDF); (3) take the square root of this value

$$RMS(e_q) \stackrel{\text{def}}{=} \left[\int_{-\infty}^{\infty} e_q^2 \cdot p(e_q) \cdot de_q \right]^{\frac{1}{2}} \tag{5.4}$$

Since the probability is not defined outside the range of $\pm q/2$, we can set the limits of the integral. Furthermore, since the PDF, $p(e_q)=1/q$, is a constant, it can be pulled out of the integral:

5. Quantization

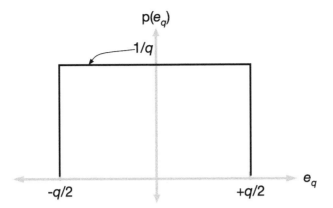

Figure 5.7

The quantization error, e_q, can be 0 if the analog level falls identically on a quantization level, or as far off as $\pm q/2$ if the analog level is exactly between two quantization levels. The probability of these particular values of e_q, or any other level for that matter, is uniform and is equal to $1/q$.

$$RMS\left(e_q\right)=\left[\frac{1}{q}\int_{-q/2}^{q/2} e_q^2 \cdot de_q\right]^{1/2} \tag{5.5}$$

After evaluating the integral and applying the limits, we find that the average noise is equal to the quantization step size divided by the square root of 12:

$$RMS\left(e_q\right)=\frac{q}{\sqrt{12}} \tag{5.6}$$

5.4.4 Programming example: PDF and RMS of quantization error

In this example, we will first plot the PDF for e_q. To start, we will generate a 100,000-point random sequence using **rand()**. Next, we will quantize this signal to 8-bits and generate the error signal by differencing the 8-bit version from the original. Since we are still in a range of ± 1 and we are using 8-bit encoding, our quantization step size does not change from the previous example. Therefore, we can predict the limits of the PDF as $\pm q/2$ (± 0.00390625) and the height of $p(e_q)$ as $1/q$ (128). The PDF will be approximated with a 50-point histogram with box edges corresponding to the limits of the error signal, e_q. Here, the 'pdf' normalization parameter affects the vertical axis scaling by setting the height of each bar to the number of observations in a bin divided by the total number of observations times the width of the bin, such that all box areas sum to one.

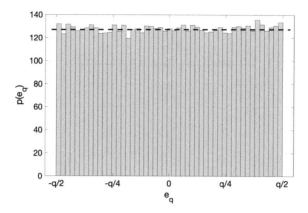

Figure 5.8

The PDF of e_q, which ranges from $-q/2$ to $+q/2$, is uniform (rectangular), with probability equal to $1/q$, or 128 for our 8-bit example.

```
% draw the quantization error pdf
n   = 2*rand(100000,1)-1.0;
n8 = quantBits(n,8,1);
eq_n8 = n - n8;

q=2A/2^N;
% for Matlab use:
histogram(eq_n8, linspace(-q/2, q/2, 50), 'Normalization', 'pdf');
% for Octave use:
hist(eq_n8, linspace(-q/2, q/2, 50);

grid on;
```

Looking at the shape of the PDF (Figure 5.8), it is clearly characterizable as rectangular. Common names for this distribution include *uniform, boxcar,* and *rectangular.* Note the average height is 128 ($1/q$) and the x-axis limits are ±0.0039 ($±q/2$).

Next, let's calculate the predicted and actual RMS level of e_q. The predicted level is q over the square root of 12, while the actual level is obtained by taking the square root of the mean of the error signal squared. In both cases we obtain the value of 0.00226 – so theory matches practice.

```
eq_rms = q/sqrt(12)   % estimate of eq RMS
sqrt(mean(eq_n8.^2)) % actual eq RMS
```

5.5 Pulse code modulation

Pulse code modulation (PCM) is a way of representing the digital audio stream on a physical medium such as a magnetic or optical disk. The conversion from a digital

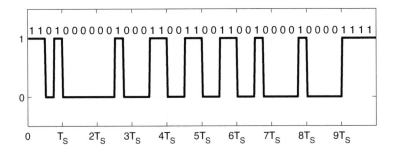

Figure 5.9

A PCM sequence for the digital signal in Table 5.1 and Figure 5.2 is shown.

binary memory buffer to PCM (and back) is completely lossless. Under PCM, a sample (which comprises N_{bits} binary digits) is represented by a series of N_{bits} modulated code pulses. For multichannel applications, the data for each channel are interleaved. The term linear PCM (or LPCM) is used when a uniform quantizer is used, such as the quantizers discussed thus far in this chapter. LPCM is common for encoding digital audio, and is used in formats such as WAV, AIFF, AES3, Blu-ray, among others. Consider the digital signal in Table 5.1 and Figure 5.2. The PCM representation of the digital audio stream is shown in Figure 5.9. Note that each sample is sub-divided into 4 pulses, with one pulse per bit.

5.5.1 Non-uniform quantization

While we want to have a large dynamic range in order to capture very small signals as well as very large signals, by virtue of the fact that all audio signals are constantly crossing zero, the quantization levels at and around zero are utilized much more frequently than the quantization levels near full-scale. In fact, it is not uncommon to have a digital headroom of −12 dB or more as an assurance against *clipping*, a harsh distortion that occurs when the signal exceeds the upper limits of the dynamic range than the converter is capable of handling. However, the irony of this scenario is that while for high-level signals, the quantization step size is minuscule relative to the signal level, for low-level signals, the quantization step size is relatively much larger. In short, low-level signals have increased relative quantization error, but are the most frequently utilized.

One approach to address this discrepancy is to employ *non-uniform quantization*, in which the quantization step size varies depending on the signal level. Under this scheme, the quantization step size is finer at low-levels (compared to linear quantization) and coarser at high-levels. Therefore, instead of preserving the step size across the dynamic range, the coarseness relative to the signal level is being preserved, as shown in Figure 5.10. As a result, the RMS of the quantization error increases for high-level signals and decreases for low-level signals. Interestingly, this coincides with the functioning of the human auditory system, in which high-level sounds have a greater ability to mask noise compared to low-level sounds.

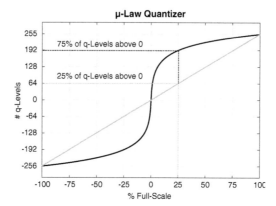

µ-Law Quantizer

Figure 5.10

Input levels (horizontal axis) and quantization index (vertical index) are shown in an input-output graph. A linear quantizer (gray) has the same step size at low- and high-level amplitudes, while a non-linear quantizer (black) has small step sizes at low input amplitudes and large step sizes at high input amplitudes.

One such non-uniform quantization is known as a *µ-law* algorithm. The µ-law algorithm distributes quantization levels on the encoder (ADC) for a signal $-1 \leq x \leq +1$, according to:

$$y = Q_\mu(x) = \text{sgn}(x) \cdot \frac{\ln\left(1 + \mu \cdot |x|\right)}{\ln(1 + \mu)} \tag{5.7}$$

Typically, µ-law quantizers are used only on low bit-depth systems, such as telephony. For example, with an 8-bit converter, the constant µ is 255 (in the United States and Japan)[1]. But a signal that has been digitized with a µ-law quantizer must be reconstructed with a DAC that utilizes an inverse of the µ-law, given as:

$$x = Q_\mu^{-1}(y) = \text{sgn}(y) \cdot \left(\frac{1}{\mu}\right) \cdot \left((1+\mu)^{|y|-1}\right) \tag{5.8}$$

5.5.2 Programming example: µ-law quantization

In the following example, you will quantize a flute solo to 8-bit using both uniform and µ-law quantization schemes. You should notice that while the µ-law quantized signal still sounds fuzzy, especially in the louder portions, the noise floor is noticeably lower than 8-bit linear quantization. The audio for this example can be downloaded at DigitalAudioTheory.com.

```
u=255;         % mu-law compression factor
[y24, fs]=audioread('flute.wav', [1 441000]);    % 10s of flute solo

% linear quantizer
y8 = quantBits(y24,8,1);
```

```
% mu-law quantizer
y24mu = sign(y24).*log(1+u*abs(y24))/log(1+u);
y8mu_comp = quantBits(y24mu,8,1);
y8mu = sign(y8mu_comp)*(1/u).*((1+u).^(abs(y8mu_comp))-1);

sound([y24; y8; y8mu],fs)
```

5.6 Challenges

1. How many quantization levels are achieved with a 10-bit ADC, and 16-bit?
2. You are digitizing an analog signal and you need to have a quantization step size of <100 μV. What is the bit-depth required for the ADC for capturing a dynamic range of -10 to $+10$ V?
3. For the ADC used in Problem #2, what is the actual achieved resolution?
4. Using the ADC in Problems #2 and #3, what new dynamic range could be achieved if we allow the resolution to increase to 150 μV?
5. How many bytes of memory must be reserved for a buffer to capture one minute of mono audio at a sample rate of 44,100 samples/second at a bit-depth of 16 bits?
6. What is the SNR in dB of a signal level that is 100x the level of the noise?
7. A microphone produces a maximum amplitude of ±4 mV and an RMS value of 2mV. The microphone signal passes through an amplifier with a gain of 40 dB prior to signal capture. You are asked to choose an ADC for this project, and a post-gain resolution of 1 mV is desired. What is the minimum bit-depth for the ADC?

   Suppose you choose an ADC with an input of ±5 V and a bit-depth of 16 bits. What is the actual (post-gain) resolution of the signal?
8. A 16-bit ADC has a ± 1 V input range. The pre-amplifier provides only an output range of ± 100 mV.
   a. What is the ADC resolution?
   b. What is the RMS of the quantization error?
   c. What is the SER (in dB) of the ratio of the peak **signal level** to the quantization error?
   d. How many bits on the ADC are being "wasted" (unused)?
9. For an ADC with quantization step-size of q, sketch the probability density function of the quantization error. Label the axes, including upper and lower bounds for e_q (horizontal axis) and $p(e_q)$ (vertical axis).
10. Solve the integral to show that the RMS(e_q) is $q/\sqrt{12}$

Bibliography

[1]   International Telecommunication Union, *Pulse Code Modulation (PCM) of Voice Frequencies*. ITU-T G.711:1993.

6

Dither

All digital audio, whether derived from analog capture or undergoing word-length reduction to final format for digital distribution (often 16 bit), experiences some amount of signal truncation. Recall from Chapter 5, a reduction of bit-depth indicates a loss of signal amplitude resolution. This truncation process introduces some artifacts that impact the sonic fidelity. First, the quantization process increases the noise floor, and second, a specific type of distortion is created, in which the noise floor modulates with (or is correlated with) the signal. This is a perceptible distortion, especially for low-amplitude signals, but fortunately, some clever engineers have discovered ways of mitigating it through the use of *dither*.

A dither signal is simply some very low amplitude noise that can be broadband or frequency-specific, which is *added* to the audio signal prior to quantization or word-length reduction. Additionally, researchers have found ways of actually "hiding" dither into frequency regions that are imperceptible to the ear! While many types of dither exist, the penultimate goal in each case is the faithful representation of the source material, with only negligible artifacts arising from the quantization process. The difference between the various dither signals is their criteria measure. Some dithers are designed to minimize the noise floor in the speech frequencies, while others are designed to impact the fewest number of LSBs; many other strategies (and, therefore, types of dither) exist.

At first blush, it may seem counter-intuitive to *add* noise to the audio signal – this appears to violate our principle of transparently representing the analog signal. However, it was discovered that the audibility of adding some very low-level noise was much lower compared to the audibility of distortions related to quantization. In the next sections, we will look at how this specific type of distortion arises, and how it is resolved through the use of additive dither.

6.1 Signal-to-Error Ratio (SER)

We already discussed SNR, which is the signal RMS level compared to the RMS level of noise, expressed as a logarithm with units of dB. For example, if we measure an audio signal to have a level of 1 V_{RMS}, then re-measure the channel with the signal removed, we might measure a noise level of 0.001 V_{RMS}. So, the SNR would then be

$$SNR = 20\log\frac{A_{signal}}{A_{noise}} = 20\log\frac{1V}{0.001V} = 60dB \tag{6.1}$$

When expressing the level of the signal compared to the level of quantization error, the concept is identical, but we change the terminology to SER. We already determined the level of the quantization error, so to determine the SER, we must determine the RMS level of the signal. For a zero-mean, bi-polar, and quasi-periodic signal, we can estimate the RMS as the peak amplitude divided by the square root of 2, which applies in the case of audio signals.

$$RMS(x) = \frac{A}{\sqrt{2}} \tag{6.2}$$

For example, let's consider the SER for a signal with dynamic range from $-A$ to $+A$ that is quantized with a bit-depth of 10 bits. To estimate the RMS level of the quantization error, the quantization step size must first be determined, by

$$q = \frac{A^+ - A^-}{2^{10}} = \frac{2A}{1024} \tag{6.3}$$

Then the RMS of quantization error is given by

$$RMS(e_q) = \frac{q}{\sqrt{12}} = \frac{2A}{1024\sqrt{12}} = \frac{A}{512\sqrt{12}} \tag{6.4}$$

And finally, we can calculate SER using the ratio of signal level to the e_q level

$$SER = 20\log_{10}\left(\frac{\left(\frac{A}{\sqrt{2}}\right)}{\left(\frac{A}{512 \cdot \sqrt{12}}\right)}\right) = 20\log_{10}\left(\frac{512\sqrt{12}}{\sqrt{2}}\right) = 61.96 \ dB \tag{6.5}$$

6.1.1 The case of the missing noise

When we calculate the SER of 61.96, the enterprising student might notice that this does **not** equal 6.02 dB × 10 bits = 60.20 dB. The "shortcut" discussed previously was that a bit adds 6.02 dB to the dynamic range of the digital signal. But in fact, there is a difference of 1.76 dB between the 61.96 dB and 60.20 dB calculations of SER. Fortunately, we can explain the origin of this number. Assuming a signal that is bi-polar and zero-mean, the peak positive and negative amplitudes (+A and −A) will have the same magnitude, meaning we can write the quantization step-size equation as

$$q = \frac{2A}{2^{N_{bits}}} \tag{6.6}$$

Which can be rewritten in terms of A as

$$A = \frac{q2^{N_{bits}}}{2} \tag{6.7}$$

Therefore, for an arbitrary sinusoid, s with amplitude A, we can estimate its RMS level as

$$RMS(x) = \frac{q2^{N_{bits}}}{2} \cdot \frac{1}{\sqrt{2}} \tag{6.8}$$

Then the SER, which is a ratio of the signal RMS to the quantization error RMS, becomes

$$SER = 20\log_{10}\left(\frac{\left(\frac{q2^{N_{bits}}}{2\sqrt{2}}\right)}{\left(\frac{q}{\sqrt{12}}\right)}\right) \tag{6.9a}$$

$$= 20\log_{10}\left(\frac{2^{N_{bits}}\sqrt{12}}{2\sqrt{2}}\right) \tag{6.9b}$$

$$= 20\log_{10}\left(2^{N_{bits}}\right) + 20\log_{10}\left(\sqrt{\frac{3}{2}}\right) \tag{6.9c}$$

$$= 6.02N_{bits} + 1.76 \; dB \tag{6.9d}$$

Looking back at the previous example, we could have quickly calculated the SER for a 10-bit converter with this equation

$$SER_{10-bit} = 6.02 \cdot 10 + 1.76 = 61.78 \tag{6.10}$$

6.2 SER at low signal levels

While the SER at high signal levels is great enough to mask the quantization error, this is not the case at low signal levels. Consider a low-level signal, x_{LL}, with an amplitude that is at −60 dB below full-scale; certainly quiet, but not unreasonably so. The amplitude of this signal is 1/1,000 of full-scale, and the RMS of this signal is

$$RMS(x_{LL}) = \frac{0.001 \cdot A_{FS}}{\sqrt{2}} \tag{6.11}$$

However, even though the signal level is low, the level of quantization error does not change. For a N-bit converter, the RMS level of e_q is

$$RMS(e_q) = \frac{2A_{FS}}{2^{N_{bits}}\sqrt{12}} \tag{6.12}$$

For the 10-bit converter from the previous example, the SER will decrease dramatically

$$SER = 20\log_{10}\left(\frac{0.001 \cdot \dfrac{A_{FS}}{\sqrt{2}}}{\dfrac{2A_{FS}}{2^{10} \cdot \sqrt{12}}}\right) = 1.97 \ dB \tag{6.13}$$

Compounding the fact that the SER is decreased at low levels, is the fact that quantization error becomes *correlated* with the signal at low levels. This is because, at low levels, several samples may pass before a signal "jumps" to the next quantization level; in other words, the signal gets "stuck" at a fixed quantization level. As a result, the quantization *error* closely tracks the phase of the signal, resulting in an e_q with frequency content that matches that of the signal. This is significant and audible during fade-outs and envelope tails, known as *noise modulation*.

6.3 Applying dither

One approach to de-coupling quantization error from the signal is to add a very small amount of additive white noise, known as *dither*. The dither signal is randomly generated and has an amplitude of half of a quantization level, $q/2$. Dither allows the quantized signal to "break out" of the quantization level at which it is currently "stuck". Since dither is randomly distributed, even small changes in the underlying signal are enough to bias the dither towards the "next" quantization level. Ken Pohlmann, in *Principles of Digital Audio*, shares the following analogy for dither [1]

> ...[O]ne of the earliest [applications] of dither came in World War II. Airplane bombers used mechanical computers to perform navigation and bomb trajectory calculations. Curiously, these computers (boxes filled with hundreds of gears and cogs) performed more accurately when flying on board the aircraft, and less well on ground. Engineers realized that the vibration from the aircraft reduced the error from sticky

moving parts. Instead of moving in short jerks, they moved more continuously. Small vibrating motors were built into the computers, and their vibration was called dither from the Middle English verb "didderen" meaning "to tremble".

The simplest form of dither derives from a pseudo-random number generator that is scaled to a level of $\pm q/2$, with all values having uniform likelihood; this is known as Rectangular PDF (or commonly RPDF) dither. Since its amplitude does not exceed more than half the quantization level, this type of dither signal cannot impact more than one quantization level. There are two instances when dither is generally added. First, on analog capture, in which case it is added *prior to* quantization (see Figure 6.1). In this case, an infinitesimally resolute analog signal is being truncated during quantization, and the additive dither helps spread the value among the two closest quantization levels. In the second case, digital dither is applied when performing word-length reduction, such as when bouncing to disk or mastering. Typically, this might involve the truncation of the 8 LSBs of a 24-bit audio signal down to 16-bit. In this case, Figure 6.1 still applies, except that the input signal and the dither are both already digital, and the quantizer is performing word-length reduction.

Like many engineering trade-offs, dither comes at a cost – since dither has an amplitude of $q/2$, we can think of it as a signal that occupies half of a bit. If you recall from earlier in the Chapter, we can estimate the impact that a bit has on the SER, as $6.02N_{bits}$. The impact of a half-bit can be determined in the same way, such that by adding dither, we are worsening our SER by:

$$6.02 \cdot 0.5 = 20\log_{10}\left(\frac{1}{2^{0.5}}\right) = -3.01dB \tag{6.14}$$

Alternatively, we can consider the amount of noise that is being added to the signal. Since RPDF dither (v_{RPDF}) has a uniform PDF, its RMS calculation is very similar to the quantization error RMS from Chapter 5, and is given by:

$$RMS(v_{RPDF}) = \frac{q}{\sqrt{12}} \tag{6.15}$$

Since the RMS of quantization error and RPDF dither are identical, then when we sum them, the increase in noise floor must be +3.01 dB. Whether thought of in this

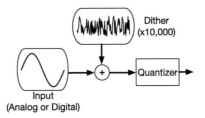

Figure 6.1

A signal is combined with dither prior to quantization. This can occur during the ADC process or in digital word-length reduction (e.g., 24-bit to 16-bit).

manner or by "losing" a half-bit, either way, it is clear that adding dither is decreasing our dynamic range (slightly) and increasing our noise floor (slightly).

6.3.1 Programming example: RPDF dither

In this example, we will create dither and apply it to a quantized signal. Then we can observe the impact on the quantization error and noise floor. We will generate 50 ms of a 40 Hz cosine with a sample rate of 40 kHz and quantize it to 8 bits using the same **quantBits()** function that was developed in Chapter 5.

```
N=8;              % bit depth
A=1;              % max amplitude
fs=40000;         % sample rate (Hz)
f=40;             % frequency (Hz)
T=0.05;           % duration (s)

x = 0.95*cos(2*pi*f*[1:T*fs]/fs);
x8 = quantBits(x,N,A);

plot(x8);
grid on;
```

Next, we will generate dither to add to the quantized signal. The dither will derive from a pseudo-random noise generator within the range of ±0.5 and a length of 2,000 samples (2,000 samples = 50 ms × 40,000 samples/s), which will be scaled by q. The dither will be added **prior to** quantizing. A comparison of an 8-bit version both with and without dither is shown in Figure 6.2.

```
q=2*A/2^N;
d=q*(rand(1,T*fs)-0.5);     % random noise in the range of +/-q/2
x8d = quantBits(x+d,N,A);   % add to the signal *before* quantizing
hold on; plot(x8d)
```

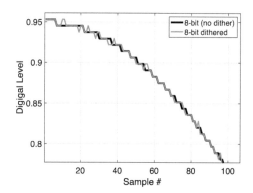

Figure 6.2

The 8-bit quantized signal (black) has a staircase shape, while the quantized signal + dither (gray) jumps back and forth between quantization levels.

Dither increases the noise floor by a predictable amount, +3.01 dB. This prediction matches closely with the amount of noise actually added if we compare the signal error levels with dither to quantization error without dither.

```
20*log10(2^0.5)
20*log10(rms(x8d-x)/rms(x8-x))
% for Matlab use:

histogram(d, linspace(-q/2, q/2, 50), 'Normalization', 'pdf');
% for Octave use:
hist(d, linspace(-q/2, q/2, 50));
```

Furthermore, it can be shown that the PDF of the dither is indeed rectangular using the histogram function. Here we set 50 equally spaced bins between $-q/2$ and $+q/2$, as shown in Figure 6.3. If RPDF dither doesn't look rectangular, try increasing the length and re-running the random number generator.

While dither successfully decouples large level signals from the quantization noise, some correlation to the signal can still be observed, which is especially evident at very low signal levels, as shown in Figure 6.4. So, while noise modulation is being reduced with RPDF dither, some still remains. To see this effect for yourself, redo the previous programming example with a very low-level signal. For example, to generate a signal at ¼ of a quantization level, use the following code.

```
t=[1:T*fs]/fs;
x = q*cos(2*pi*f*t)/4;

% note: the 8-bit encoder *can't* capture the 0.25-bit signal
% add "rectangular" dither
x8d = quantBits(x+d,N,A);
```

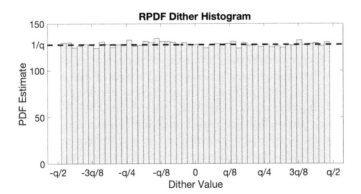

Figure 6.3

The PDF estimate of the dither signal – the height of each bar is related to the number of observations within the RPDF dither signal for that range of values (related to the width of each bar). The sum of all of the bar areas is exactly 1. Note that the average height is equal to $\frac{1}{q} = \frac{1}{0.0078125} = 128$.

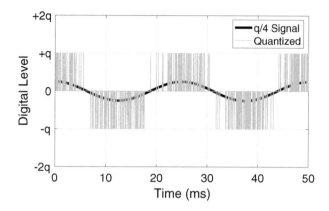

Figure 6.4

A low-level signal (black) that is less than one quantization level will not be encoded by a quantizer at all, unless dither is added (gray). Then in can clearly be seen that the quantization error (e_q) is still highly correlated with the signal.

```
plot(t,x); hold on;
ylim([-2*q 2*q]); grid on;
plot(t,x8d); grid on;
```

6.4 Triangular PDF dither

A different type of dither can be created from the rectangular dither, known as triangular PDF (or TPDF) dither. The intrepid music technology student may recognize that a triangle shape can be generated from the convolution of two rectangles (if this is not obvious to you at this point, that is okay). Furthermore, convolution of two PDFs corresponds with addition or subtraction of random variables in the time domain. This complicated process can be imagined with a pair of dice. If you roll a single die, it is well known that the probability of rolling any given number is uniform (1/6); in other words, a single die roll exhibits a RPDF distribution. However, if we roll a pair of dice, which are completely random and independent of one another, then the likelihood of rolling any *sum* of numbers is no longer uniform. There are a total of 36 possible combinations that sum to possible values from 2 to 12. There is only one throw that can produce a 2: 1+1; however, there are 6 throws to total a 7: 6+1, 1+6, 5+2, 2+5, 4+3, 3+4. The PDF is shown in Figure 6.5, exhibiting a triangular shape.

Since the dither is randomly generated any subsequent value will be independent of the preceding value. If we add or difference the dither signal with its previous value, then we can generate TPDF dither with a single memory unit and a pseudo-random number generator. It is less complicated than it sounds.

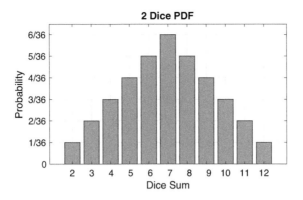

Figure 6.5

The probability of throwing a particular sum (for 2 dice) is maximum at 7, and minimum at 2 and 12. While each individual die has a rectangular PDF, the sum (or difference) of two dice has a TPDF.

6.4.1 Programming example: TPDF dither

The generation of TPDF dither uses a process known as "backwards differencing". Conceptually, a backwards differencing algorithm *accentuates* or *enhances* the changes in a signal from one moment to the next and *diminishes* any constant values from one moment to the next. Imagine a signal that changes from 0.9 to 0.85 – when these are differenced, we are left with −0.05, a small value. On the other hand, if the signal changes from 0.9 to −0.9, then the difference is −1.8, a much larger value! We will utilize backwards differencing to generate TPDF dither. Load in the low-level 40 Hz signal, sampled at 40 kHz, with RPDF dither from the 6.3.1 programming example. When TPDF dither is applied, noise modulation is no longer evident (Figure 6.6).

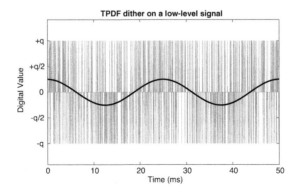

Figure 6.6

TPDF dither is decoupled from the low-level signal (1/4 quantization level).

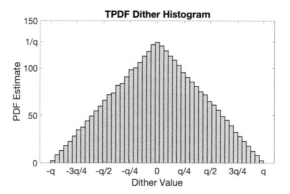

Figure 6.7

We see that the PDF is truly triangular, but the range of possible error values grows to $[-q, +q]$, so the maximum quantization error has increased.

```
d_tri = zeros(size(d));
d_tri(1) = d(1); % the first value of d_tri is the same as d

% each subsequent sample is a difference with the preceding value
d_tri(2:end) = d(2:end)-d(1:end-1);
```

Finally, we can examine the PDF estimate of the TPDF dither. Note that here the range of TPDF dither values range from $-q$ to $+q$, which is twice as large as the range for RPDF dither (Figure 6.7).

```
figure();
% for Matlab use:
histogram(d_tri, linspace(-q, q, 50), 'Normalization',
'probability');
% for Octave use:
hist(d_tri, linspace(-q, q, 50));
xlim([-q q])
```

6.4.2 Noise shaping

Differencing the rectangular dither with itself achieves decoupling from the signal but at the cost of increasing the quantization noise to the range of $-q$ to $+q$ instead of $-q/2$ to $+q/2$. While the RPDF distribution is defined simply by a constant $(1/q)$, the TPDF distribution is a piecewise definition of two slopes with offsets of $1/q$. For this reason, the calculation of the RMS is not as simple as the RPDF case, but we can look up this value [2], as it is well described for a zero-mean, symmetrical TPDF, and is given by

$$RMS(v_{TPDF}) = \frac{q}{\sqrt{6}} \tag{6.16}$$

It can be seen that the difference between RPDF and TPDF RMS levels is a factor of $1/\sqrt{2}$. In other words, TPDF adds an additional 3.01 dB to our noise floor. But this has been cleverly done due to the fact that we are differencing the rectangular dither *with itself* (as opposed to a different random signal altogether).

Think for a moment about a slow-moving average, a good example might be the two-day moving average of stock prices. The average has less fluctuations and is smoother: it could be characterized as slower-moving (i.e., lower frequency) than the day-to-day signal. The moving average is, in fact, an LPF, and its principle operator is addition (+). But if we flip the addition to subtraction (–), and we instead <u>difference</u> the current day to the previous day, then we are seeing the day-to-day fluctuation (instead of the day-to-day average). The *differenced signal* tends to stay closer to 0 since it is removing the trend. It, in fact, is attenuating the slow-changing portions of the signal and emphasizing the rapidly changing portions of the signal. Since the TPDF dither was created from differencing a single sequence from itself, then the TPDF dither has this high-pass characteristic too. This has the end result of increasing the dither noise in the high frequencies (HF) but decreasing it in the low frequencies (LF). This perfectly complements the human ear's natural abilities – the ear is much less sensitive to HFs. So, while TPDF dither actually has more noise than RPDF dither, it perceptually *sounds* quieter!

In Chapter 1, an introduction to *frequency response* was given that drew an analogy to a parametric equalizer (EQ) – we can imagine tunable frequency knobs being mapped to the horizontal axis and gain settings to the vertical axis. With this analogy, a gain at a certain frequency results in a value above 0 dB, while a value below 0 dB indicates an attenuation at that frequency. In Figure 6.8, the frequency responses for RPDF and TPDF dither are shown. The takeaway here is that while triangular dither has more gain in HF, it has more attenuation in low- and mid-frequencies, making this type of dither more pleasing to the ear, even though it adds more noise overall.

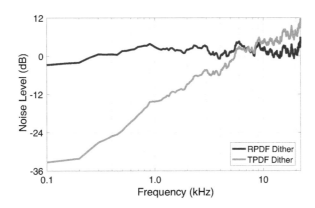

Figure 6.8

Even though TPDF dither (gray) has more energy than RPDF dither (black) overall, it is concentrated in the HF, where the human ear is much less sensitive.

6.5 High-frequency dither

With RPDF dither, the dither signal itself is generated from a pseudo-random noise generator, in which every frequency is represented in equal amounts. We saw with TPDF dither, that the energy was shifted to the HF, a frequency region where the ear is less sensitive. If we follow this line of thinking to its logical conclusion, we can envision a dither signal that is purely comprised of HF – this is known as high frequency, or sub-Nyquist, dither. Depending on the sample rate, the location of this dither could change, but the idea is to locate it within a 1-kHz band, just below the Nyquist frequency. For example, if the sample rate was set to 44.1 kHz, then the highest representable frequency is half that, or 22.05 kHz – in this case, the dither would be centered at 21.55 Hz with a bandwidth of 1 kHz (meaning ±1/2 kHz).

As an alternative to generating high-frequency dither by simply filtering a wideband dither signal is to devise a deterministic algorithm, like a pseudo-random number generator, that produces a dither signal with the desired frequency and PDF characteristics. This is a difficult proposition that is considered trade secret among manufacturers who have devised good high-frequency dither sequences. Among the most popular is the Apogee UV22 dither box and plugin (UV for "ultra violet", indicating HF, and 22 as a nod to the dither frequency, at or around 22 kHz). Another popular high-frequency dither was made by the POW-R Consortium – this is a group of audio companies who are interested in collectively developing an industry-standard dither signal. The POW-R dither algorithms, which are licensed to most leading DAW vendors, place most of the dither in the HFs (near Nyquist), but also create EQ notches in frequency regions that are particularly sensitive to the ear – for example, the POW-r3 algorithm creates a dither signal with a pronounced dip in the 3 kHz to 4 kHz region, where the ear is highly tuned to these speech-frequency regions. Figure 6.9 shows some examples of these types of dither. The type with the lowest RMS is POW-r2 (at −89 dB for 16-bit). The UV22HR dither signal has clear periodicity, resulting in a large spike in the frequency response near 22 kHz. The most perceptually shaped, but also with the highest noise floor (−77 dB for 16 bit), is POW-r3, which spans more than 20 quantization levels.

The question arises about the appropriateness of each type of dither. Certainly, dither should always be used on ADC analog capture, as well as on any word-length reduction (for example from 24-bit down to 16-bit). It is common for a mastering engineer to apply a psychoacoustically shaped dither signal for the final formats (typically in 16 bits) – for example, the POW-r3 dither signal. When possible, the delivery to a mastering house should be at 20-bit resolution, or greater. But if dithering must be applied prior to final format, then the reasonable approach is to use the dither with the lowest noise floor – this prevents a build-up of dither noise in the case that more than two dither signals are applied. Finally, if "freezing" a track (to free up the processors), dithering should be applied, and in this case, a low noise floor dither is indicated.

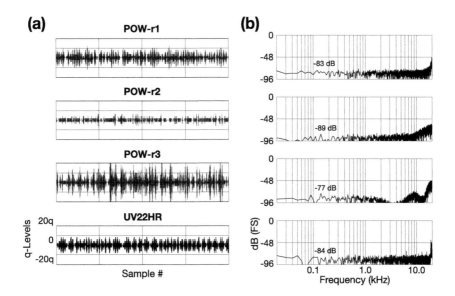

Figure 6.9

The dither signals (left) and their frequency responses (right) are shown for three types of POW-r and for UV22HR. All of these have a slope towards the HF where the ear is the least sensitive. They also demonstrate varying amplitudes, from ±8 quantization levels to ±38 quantization levels.

6.6 Challenges

1. Compute the SER for a 24-bit audio signal with no dither.
2. For Challenge #1, what is the new noise floor (in dB-FS) when RPDF dither is added?
3. What is the maximum number of quantization levels (or LSBs) that a TPDF dither can impact?
4. Prove that TPDF dither has an RMS value of $q / \sqrt{6}$, given that the PDF of a triangular shape with bounds of $\pm q$ is defined by

$$p_{TPDF}(x) = \begin{cases} \dfrac{1}{q} + \dfrac{x}{q^2}, & x < 0 \\[2ex] \dfrac{1}{q} - \dfrac{x}{q^2}, & x \geq 0 \end{cases}$$

Recall the definition for RMS is given by

$$RMS(x) \overset{\text{def}}{=} \left[\int_{-q}^{q} x^2 \cdot p(x) \cdot dx \right]^{1/2}$$

6.7 Project – dither effects

Introduction

In this project we will be exploring the concepts of quantization and dither using the function **quantBits**(), which takes three arguments: (1) the audio signal being quantized, (2) the bit-depth for quantization, and (3) the full-scale level. Just like in the programming examples in this chapter, we will define a pseudo-random noise signal with the **rand**() function. (<u>Note</u>: There are a few random sequence generators to choose from. In addition to **rand** are, **randi**(), which generates random integers, and **randn**(), which generates random values in a Gaussian/normal distribution, meaning it has no theoretical maximum value.) **rand**() generates random uniformly distributed real numbers in the range of 0 to 1. In general, you can generate N random numbers in the interval [a, b] with the following formula:

```
r = a + (b-a).*rand(N,1);    % or (N,2) for stereo
```

One command that will make your life easier is **linspace**() – this command will give you a vector starting with some starting value, ending in another, and specified length. For example, we can generate a vector with a specified number of elements using

```
vector = linspace(startValue, endValue, N);
```

Assignment

    a. Generate a 500 Hz sine wave of 3 seconds in length with an amplitude of ±1. Use **quantBits**() to quantize it to 8, 4, and 2 bits, respectively. Describe what you hear for each bit-depth in your report.

    b. Plot a few periods of the waveform in a 4 × 1 subplot showing how the waveform changes at each bit-depth.

    c. Use **quantBits** to quantize the signal from part a to only 1 bit. Plot the original and quantized versions in a 2 × 1 subplot.

    d. Generate a 500 Hz sinusoid 3 seconds in length with an amplitude ±0.4 and use **quantBits** to quantize the signal to only 1 bit. Plot the original and quantized versions in a 2 × 1 subplot.

    e. Generate a noise signal using **rand**() that has an amplitude of ±0.5. Add it to the *unquantized* signal from part d.

    f. Use **quantBits** to quantize the *dithered* signal from part e to only 1 bit. Plot the original and quantized versions in a 2 × 1 subplot.

    g. Listen to all six versions of the signal and describe what you hear for each. For the signal in part f, what <u>two</u> benefits did we achieve by adding noise to the signal?

    h. Repeat parts **e.** and **f.** using dither that has a triangular probability distribution function. <u>Hint</u>: the amplitude of the TPDF should be 1.

Bibliography

[1] Pohlmann, K., *Principles of Digital Audio*, McGraw-Hill, New York, 2011, pp. 16–30.

[2] Wannamaker, R., "The Mathematical Theory of Dithered Quantization," Ph.D. Dissertation, Dept. of Applied Mathematics, Univ. of Waterloo, ON, Canada, (1997 Jun.).

7

DSP basics

Digital Signal Processing (DSP) is a means of operating on a digital signal in order to effect it in some way. These digital effects (or sometimes *fx* for short) alter the signal in a way such that upon reconstruction and reproduction, the signal has a different frequency, timing, or magnitude properties. In digital audio signal processing effects such as EQ, amplitude compression and expansion, time-based contraction and expansion, echo, phase effects, waveshaping, etc. are often included, among many others. One way of characterizing these is into two broad categories: linear and nonlinear. For a review on linearity, see Section 1.3.2 in Chapter 1. In Chapter 7 and the next several, linear effects will be discussed, and then Chapter 14.5 includes an Introduction to non-linear system characterization.

DSP is implemented programmatically in software (perhaps in the form of a DAW plugin) or firmware (for example, found in many audio amplifiers and pedals on an embedded platform). The backbone of a DSP effect is known as the *block diagram*, which is a graphical representation of all of the operators that affect the signal as it passes through the effect. For linear systems, these comprise only time-shift operators such as delays and advances, multiplication operators such as gains and attenuators, and finally accumulators (also known as summers). In this chapter, we will explore each of these operators and start piecing them together into block diagrams. Finally, we will convert these block diagrams into equations that can be implemented in code.

7.1 Time-shift operators

In Chapter 3.3, the time-shift operator was introduced, which transports either past samples or <u>future</u> samples in order to process them with the <u>current</u> sample. These operators are known as *delays* and *advances*, respectively. In real-time processing, which is typical in audio, it may seem counter-intuitive to have prior knowledge about what a future sample might be; however, for certain types of filters, these future samples are required (these will be covered in Chapter 8). There is a simple workaround, though, to give a filter access to what it believes to be future samples, and that is simply to delay the entire input by some amount. For example, if a filter requires access to 10 ms of future samples and if we buffer our input by this amount (that would be 441 samples at f_S = 44.1 kHz), we can then treat these buffered samples as "future" samples – this process is diagrammed in Figure 7.1.

In this text, the input signal is denoted by x while the sample index is denoted by n; therefore, the current sample is accessed as $x[n]$. Past samples will have occurred previously, so they will be accessed by recalling previous samples, $x[n - k]$, where k is the delay amount. We can also use k as an advance amount, but since they are occurring at a later sample index, they are accessed as $x[n + k]$. For example, consider a signal $x[n]$ = [1.0, −0.2, 0.1, 0.0]. This signal is plotted in Figure 7.2 both delayed by three samples, $x[n - 3]$ (diamonds), as well as advanced by three samples, $x[n + 3]$ (circles). It may seem counter-intuitive for a signal with an index of $[n - 3]$ to start at +3, rather than −3. But the correct way to interpret the delay is to think, "what must n be in order for $n - 3$ to evaluate to 0?" (where 0 represents the first index of the signal x).

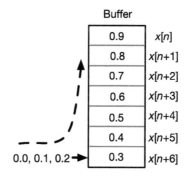

Buffer

0.9	$x[n]$
0.8	$x[n+1]$
0.7	$x[n+2]$
0.6	$x[n+3]$
0.5	$x[n+4]$
0.4	$x[n+5]$
0.3	$x[n+6]$

0.0, 0.1, 0.2 →

Figure 7.1

Consider a signal x = [0.9, 0.8, 0.7, ..., 0.0] being fed into a buffer. Samples are fed into the bottom and are "pushed" to the top. If a filter reads from the top of the buffer ($x[n]$), then it has access to what it can consider as future samples (here, up to 6 future samples). However, this comes at a cost of the signal being delayed. Delay amounts must be carefully considered since excessive delays (>50 ms) can be deleterious during real-time playback.

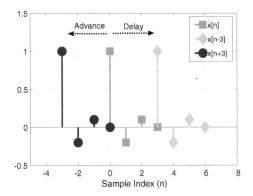

Figure 7.2

A signal, x[n] (squares), shown delayed by three samples, x[n – 3] (diamonds), and advanced by three samples, x[n + 3] (circles).

7.2 Time-reversal operator

A time-reversal operator "flips" the signal such that its last value becomes its first value (and vice versa). This operation requires access to future samples, just like the time advance operator. Therefore, a time-reversal requires the use of a buffer (such as the one in Figure 7.1) to have access to "future" samples. But this comes at the expense of delaying the output by some amount. A time-reversal is achieved by negating the index, whereby $x[n]$ becomes $x[-n]$, this in effect flips the direction in which the samples are processed. But a time-reversed signal can also be delayed or advanced. For example, a reverse-delay is $x[-(n - k)] = x[k - n]$, while a reverse-advance is $x[-(n + k)] = x[-k-n]$. A short signal, $x = \{1.0, -0.2, 0.1, 0.0\}$, is shown in Figure 7.3 both time-reversed (squares) and reverse-delayed (diamonds).

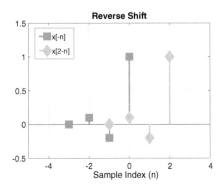

Figure 7.3

A reversed signal, x[–n] (squares), shown delayed by two samples, x[2 – n] (diamonds). Note that delayed signals, even when time-reversed, still shift in the positive n direction.

Table 7.1 A Sequence $x = [1.0, -0.2, 0.1, 0.0]$ Is Time-Reversed and -Delayed by Two Samples.

n=	2 − n=	x[2 − n]=
−2	4	$x[4] = $ (n.d.) 0
−1	3	$x[3] = 0.0$
0	2	$x[2] = 0.1$
1	1	$x[1] = -0.2$
2	0	$x[0] = 1.0$
3	−1	$x[-1] = $ (n.d.) 0

The location, n, of the samples can also be determined answering the question, "For any given sample index, n, what is the index of x that is being accessed?" Consider again the reverse-delayed signal $x[2 - n]$ in Figure 7.3, specifically sample index $n = 2$. Here, the index of x that is being accessed is: $x\left[\underset{k}{2} - \underset{n}{2}\right] = x[0] = 1.0$. In Table 7.1, additional sample locations are shown, and it can be seen that the sample values in the right-most column match the sequence shown in Figure 7.3. In instances where a signal is not defined (n.d.) for a specific index, we will assume a value of zero.

7.3 Time scaling

In the previous sections, the index, n, has been added to (in order to advance the signal), subtracted from (in order to delay the signal), or negated (in order to time reverse the signal). But if n is multiplied by a non-unity value, then *time scaling* is achieved. If this scale factor is less than one, then we achieve time expansion, and for values greater than one, time contraction. For example, consider the same sequence x, as above. If we apply a time scaling factor of 0.5, then we achieve a new sequence that is eight samples long, compared to the original four samples, as shown in Table 7.2. As in previous cases, when a non-defined index is attempted to be accessed, then a value of zero must be inferred.

Table 7.2 The Same Sequence, x, is Time Scaled by a Factor of 0.5, Resulting in the Length of the Signal Being Expanded by a Factor of 2.

n =	0.5n=	x[0.5n]=
0	0	$x[0] = 1.0$
1	0.5	$x[0.5] = $ (n.d.) 0
2	1	$x[1] = -0.2$
3	1.5	$x[1.5] = $ (n.d.) 0
4	2	$x[2] = 0.1$
5	2.5	$x[2.5] = $ (n.d.) 0
6	3	$x[3] = 0.0$
7	3.5	$x[3.5] = $ (n.d.) 0

Table 7.2 shows that creating new indexes in between samples results in discontinuities being introduced between the original samples. Therefore, with time expansion, interpolation is required. We previously used interpolation to infer the value of a signal in-between sample locations back in Chapter 4.4. We can utilize the same concept to infer a new sample value in-between two known sample values. The simplest (and probably worst) way to do this would be a linear interpolation between two samples – this can *miss* high-frequency content that is actually recoverable. A better approach would be a sinc() interpolator, like the one used in the digital-to-audio reconstruction. Whatever the approach, time expansion does require interpolation to fill in missing values.

Time contraction suffers from the opposite problem. If we apply a time scale of 2, then the signal becomes shortened by a factor of two. (To see for yourself, try replicating Table 7.2, but with a middle column of "$2n$" instead). This process introduces a different type of distortion, though. If some number of samples in a sequence are being excluded, then that signal is losing some of its sampling resolutions. Since HFs need more sample density in order to be captured discarding samples results in a loss of HFs. This process will result in aliasing, unless the signal first flows through an anti-aliasing LPF <u>prior to</u> time scaling.

7.4 Block diagrams

Block diagrams are visual representations of the processes being performed on a signal, including time-shift operations, splitting a signal, applying gain, and summing signals together. In fact, a typical block diagram will contain each of these fundamental elements. In this text, the left side of the block diagram contains the input of a signal, x, while the right side produces the output, y. The flow of the input through the block diagram is denoted by lines that move from left to right, and may contain arrows, indicating directionality. Where a line joins with one of the operators, that operation is performed on the signal, in the order that each sample of the signal arrives at each operator. Some audio DSP platforms are programmed with block diagrams, including Max/MSP (Cycling '74, San Francisco), Matlab® Simulink (Mathworks, Natick, MA), Pure Data (Miller Puckette), and SigmaStudio (Analog Devices, Norwood, MA).

As we saw previously, time-shift operations comprise delays and advances, which both involve some shift amount k, but with different signs, depending on the direction of shift. A delay operator can be thought of as a holding pen for samples – that is, once a sample enters, it is retained by the delay, according to the delay amount. A time-shift is graphically represented with a square box containing the variable z^k, where k is the shift amount (therefore, a delay of ten samples would be written z^{-10}), as in Figure 7.4(a). The reason for using the variable z will be explained shortly, but for now, it is fine to simply recognize this as a delay and to be able to identify the shift direction and amount.

A signal splitter is just like a copy operation on a computer – a splitter has one input, and two or more output *taps*, each with identical copies of the input. A splitter is graphically represented by a filled-in circle, with one line entering on the left, and two or more lines exiting towards the right, as in Figure 7.4(b).

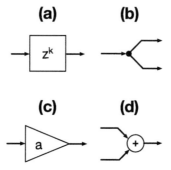

(a) **(b)**

z^k

(c) **(d)**

a +

Figure 7.4

The building blocks that illustrate the signal flow and signal operators, known as a block diagram, comprise, (a) delay, (b) splitter, (c) gain, and (d) accumulator, or summer.

The function of the gain operator is to multiply every value in a signal by the gain amount, which can be positive or negative, greater or less than one. A gain is graphically represented by a triangle, with the gain amount indicated inside the triangle, as in Figure 7.4(c).

Finally, the accumulator sums up all of the inputs entering from the left on a sample-by-sample basis and outputs the summed value. The accumulator is represented by a circle with a + or $\sum$ on the inside – see Figure 7.4(d).

7.4.1 Example: moving average filter

A *moving average filter* averages the current input with a select number of previous inputs. These are commonly employed to decrease sample-to-sample variability and to emphasize underlying trends in the signal. For example, we can expect weather during the month of July to generally increase in the Northern hemisphere. Although cloudiness or rain may cause day-to-day fluctuations in measured temperature, if we, instead, consider a moving average (perhaps one week) of temperatures then these daily fluctuations will be averaged out, and the general trend (getting hotter) will be revealed.

Consider a four-point moving average filter, which averages the current sample with the three preceding samples, as shown in Figure 7.5(a). The first output, y, is only the first input, x, scaled by 1/4 since it is averaged with 3 "preceding" values that are not defined (therefore, they must be treated as zero valued). Once the fourth sample is reached, then the output is that sample averaged with the first three. But a more efficient structure is realizable. Since this is a linear system, the position of the delays, gains, and summers can be rearranged, as shown in Figure 7.5 (b), which has three fewer delays and gains.

One important technique to note is the position of the gain. An accumulator holds digital values and is subject to the same dynamic range limitations as each sample. Therefore, if many samples are being summed together, then even if no individual sample is clipping, the sum of many samples could clip. Therefore, it is good practice to place attenuators prior to the accumulator (rather than after) to prevent clipping.

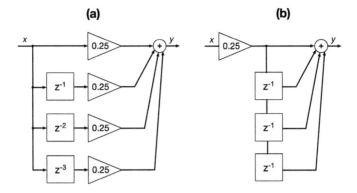

(a)   **(b)**

Figure 7.5

A four-point moving average filter produces an output, y, that is the sum of four taps comprising the current sample (top tap), the immediate previous sample (2nd tap), and so on. Previous sample values are preserved by the delays, which buffer a sample by the amount indicated (for example, z^{-3} delays a value by three clock cycles). The more efficient realization of (a) is shown in (b). Instead of delaying the current signal by 1, 2, and 3 samples, respectively, we will delay the current sample by 1 then again by 1 (to achieve a delay of 2), then one final time by 1 (to achieve a delay of 3). And instead of scaling each tap by a gain of 0.25, the input sample will be scaled by 0.25 prior to being split.

7.4.2 Delay using the z-operator

The delay operation was previously described using an operator simply called z^k, but without explanation. Here, the origin of this operator will be clarified. Consider a complex signal with frequency f and unity magnitudes for the *Re* and *Im* parts, which can simply be written in exponential form, using Euler's formula:

$$x[n] = \cos(2\pi f n / f_s) + j\sin(2\pi f n / f_s) = e^{j2\pi f n / f_s} \qquad (7.1)$$

Now, in order to delay $x[n]$ by k samples, then we simply replace n with $n - k$, which is written as:

$$x[n-k] = e^{j2\pi f(n-k)/f_s} \qquad (7.2)$$

Finally, if we factor out a term, we see that $x[k - n]$ is simply $x[n]$ times a complex exponential:

$$x[n-k] = e^{-j2\pi f k / f_s}\, \overbrace{e^{j2\pi f n / f_s}}^{x[n]} \qquad (7.3a)$$

$$= e^{-j2\pi f k / f_s} x[n] \qquad (7.3b)$$

7.4 Block diagrams

Therefore, a delay operation is simply a multiplication by a complex factor! In DSP, it is common to replace $e^{j2\pi f/f_s}$ with z, which is the reason that z^k is used to represent delay in a block diagram:

$$x[n-k] = z^{-k}x[n] \tag{7.4}$$

7.5 Difference equations

A *difference equation* is a mathematical expression of a block diagram, that is implementable in software, and is the backbone of all DSP. Difference equations are so-called because they express the relationship between the input and output, and is typically in the form

$$y[n] = a_0x[n] + a_1x[n-1] + a_2x[n-2] + \dots \tag{7.5}$$

As you can see, there is a single sample output, $y[n]$, which comprises a sum of many other values. When writing the difference equation from a block diagram, it is good practice to count the number of taps feeding the final accumulator, and leave blanks for them to fill in later. Looking back to the Moving Average filter example, since four taps are being summed together, we would start constructing the difference equation just like this:

$$y[n] = \_ + \_ + \_ + \_ \tag{7.6}$$

The next step is to determine what goes in each of those positions. Looking at the topmost tap, it is the current sample, $x[n]$ which is scaled by 0.25. The next tap is $x[n-1]$, also scaled by 0.25 (and so on). Therefore, the difference equation for the four-point moving average filter in Figure 7.5 can be written:

$$y[n] = 0.25x[n] + 0.25x[n-1] + 0.25x[n-2] + 0.25x[n-3] \tag{7.7}$$

Or equivalently:

$$y[n] = \sum_{k=0}^{3} a_k x[n-k] \tag{7.8}$$

where, a_0, a_1, a_2, and a_3 all equal 0.25.

While we can write a difference equation from a block diagram, we can also draw a block diagram from a difference equation. Consider the following difference equation:

$$y[n] = 0.5x[n] + 0.2x[n-1] + 0.1y[n-1] \tag{7.9}$$

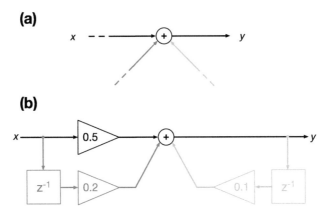

Figure 7.6

(a) Begin a block diagram with the output accumulator, leaving space for the correct number of taps; here we have three. (b) Two of the taps derive from the input (black, dark gray), while one of the taps derives from a previous output (light gray) – the previous output state is saved in a delay.

First, we should note the number of taps that are feeding the output, $y[n]$, and we can see that there are three; therefore, the block diagram will include an accumulator with three taps, as shown in Figure 7.6(a). Next, we can see that two of the taps involve x, one with the current sample (no delay, black) and a gain of 0.5, and another tap that is delayed by one sample, with a gain of 0.2, as shown (dark gray) in Figure 7.6(b). Finally, we see a y term being fed back into the accumulator, which comes from the previous output. In other words, the output must be delayed by one sample and scaled by 0.1, as shown (light gray) in Figure 7.6(b). When a copy of the output is fed back and combined with an input, this is known as a *feedback* or *recursive* tap.

7.6 Canonical form

In the examples discussed above, there is a direct realization of the gain values between the block diagrams and difference equations – this is known as *direct form*, or sometimes *direct form 1*. While the direct form is convenient for developing DSP filters, it actually is not the most efficient form. The filter's *order* is the greatest delay that exists, symbolized by N. For example, in Figure 7.6, the order is 1, since no tap has a delay greater than one. However, if we implement this either as a block diagram or as a difference equation, it is apparent that two delays will be required, one for $x[n-1]$ and one for $y[n-1]$. Theoretically, a filter should be realizable with no more delays than the highest order of the system, a form that is known as *canonical form* or sometimes *direct form 2*.

In order to convert from direct form to canonical form, the first step is to identify that there are essentially two sub-filters in cascade, one handling the inputs

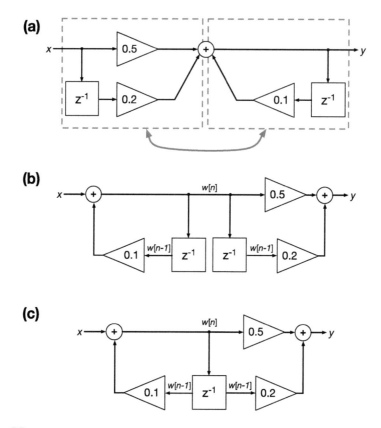

Figure 7.7

(a) The direct form filter structure can be thought of as two sub-filters in cascade. (b) The order of these sub-filters can be swapped. Note that the input into both delays is now the same value. (c) Canonical form utilizes the fewest number of delays, equivalent to the order of the filter.

and one handling the outputs, that then get summed together. And since we are considering only linear filters, for now, the ordering of these two sub-filters can be swapped, as shown in Figure 7.7(a). Once we do this swap, we need to introduce a new *state* or variable, let's call it $w[n]$, which represents an interim state of the filter. But now it can be seen in Figure 7.7(b) that the tap feeding to the two separate delays is actually the same value, and therefore the delays are each producing the same value on their respective outputs and are redundant. We can consolidate these two delays into one, thereby realizing a more efficient filter, as shown in Figure 7.7(c). In the canonical form, it is not possible to write a difference equation simply in terms of x and y, a new state variable must be utilized. In canonical form, we will have a system of difference equations, one for the output, y, and one for the state variable, w:

$$y[n] = 0.5w[n] + 0.2w[n-1] \qquad (7.10a)$$

$$w[n] = x[n] + 0.1w[n-1] \qquad (7.10b)$$

7.6.1 Programming example: filter implementation

We will implement both the direct and canonical form filters in code to verify that they indeed have the same effect on a signal. We will start by creating an impulse, and then pass this impulse through two different sets of difference equations: one that implements the direct form structure, and a set of equations that implements the canonical form. In order to preserve a delayed copy of the input and output, we will create variables called **xnm1** and **ynm1**, which will hold the values for $x[n-1]$ and $y[n-1]$ respectively, and will be initialized to 0. For canonical form, we only need one of these variables, and it is called **wnm1** (for $w[n-1]$), also initialized to 0. Note, the character sequence 'nm1' is a mnemonic for "n minus 1" that is commonly used in DSP programming. An alternative form is xn_1.

```
% create a delta input
x = [1; zeros(10,1)];

% delays for direct form
ynm1=0;
xnm1=0;

% single delay for canonical form
wnm1=0;

% execute the difference equation
for ii=1:length(x)
    % direct form equation
    y_df(ii) = 0.5*x(ii)+0.2*xnm1+0.1*ynm1;

    % set delays
    xnm1=x(ii);
    ynm1=y_df(ii);

    % canonical form equations
    w=x(ii)+0.1*wnm1;
    y_c(ii)=0.5*w+0.2*wnm1;

    % set single delay
    wnm1=w;
end
format long;
[y_df' y_c']
```

You should see identical outputs for both **y_df** and **y_c**!

7.7 Challenges

1. For the filter structure shown in Figure 7.8.

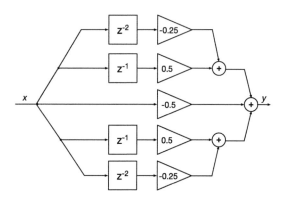

Figure 7.8

    a. Calculate the difference equation with an output accumulator with only three taps.

    b. Re-draw the block diagram using only two delays, three gains, and one accumulator.

2. Consider the following block diagram in Figure 7.9.

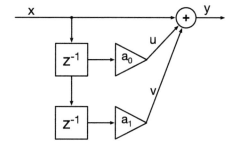

Figure 7.9

    a. What is the difference equation in terms of x, u, and v?

    b. What is the difference equation in terms of x alone?

    c. What is the order of the filter?

    d. What is the total number of delays required to implement this filter?

3. Draw the block diagram for the following difference equation:

$$y[n] = a_0 x[n] + a_1 x[n-1] + a_2 x[n-2] - b_1 y[n-1] - b_2 y[n-2]$$

4. Convert the block diagram from Question 3 from direct form to canonical form.

7.8 Project – plucked string model

Introduction

In this lab you will create a filter that implements a plucked-string model, known as Karplus-Strong algorithm [1]. It is a precursor to what would later become a generalized digital waveguide synthesis. The block diagram below describes a plucked-string model for a fundamental period L, which is related to the frequency and sampling rate by $L = f_S/f$. Additionally, there is a decay factor, d, that can be set very near to (but less than) 1.

You may notice in Figure 7.10 a moving average filter being fed back from the output to the input. But this moving average is tunable with the factor a, which sets the brightness; nearer to 1 is brighter, and nearer to 0 is muted. The input, x, is a noise burst of length L, which can be created using the **rand()** function. You might notice that the delay amount is equal to the length of the input (L). Therefore, there is never a summing interaction between x and the feedback loop. For this reason, you can read the output right out of x, and at exactly the sample index that was just read, you can store the output of the moving average filter. Then when playback reaches sample L, if we loop back to the beginning of x, we are now reading the L-sample delayed feedback.

Assignment

Implement the Karplus-Strong algorithm. You will have to choose values for L, a, and d. Play around with different values until you find something that sounds good.

1. Derive the difference equation for y in terms of L, a, and d.

2. Turn this into a function that takes three parameters: frequency, sample-rate, and brightness.

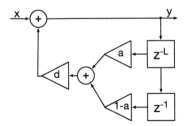

Figure 7.10

    a. Hint: You may want to have a special case for when you are reading out of the first sample of x[1], since the "previous" sample is not at x[0], but rather x[end]

3. Write a separate script that calls your function to play a chord and a melodic sequence.

Bibliography

[1]   Karplus K., Strong A. "Digital Synthesis of Plucked-String and Drum Timbres." *Computer Music Journal* Vol. 7, No. 2 (1983): 43–55.

8

FIR filters

This chapter will cover a type of filter structure known as a *finite impulse response* (FIR) filter, sometimes called a *feed-forward* filter. The concept of IRs can be reviewed in Chapter 1.3.4. But in short, an IR of a digital system is the output ("response") of the system when the input is a delta ("impulse"). Since a filter preserves previous inputs in the delay elements, even though a delta produces an instantaneous output, that delta must be "flushed" from all of the delays before a zero-valued input produces a zero-valued output. This duration is known as the *length* of the IR. As will be shown in this chapter, the length of an IR that is fed into an FIR type filter will always be finite in length. A separate class of filters, characterized by having infinitely long IRs, will be discussed in Chapter 10.

Fortunately, an FIR filter is easy to recognize! On the block diagram, an FIR filter will only have feed-forward structures – there will be no copies of the output feeding back and joining input taps. And in the difference equation, an FIR filter will only have delayed versions of the input ($x[n]$) and no delayed versions of the output, in the form $y[n]=a_0x[n]+a_1x[n-1]+a_2x[n-2]+\ldots$. By only utilizing previous inputs with no feed-back structures, the FIR filter will necessarily have an IR whose length (in samples, n) is determined by the highest number of taps (also known as the filter *order*). Other advantages of FIR filters include guaranteed stability (refer back to Section 1.3.4), ability to handle a large number of taps, and they can be designed to have no phase distortion (known as linear phase) – all these topics will be discussed in this chapter.

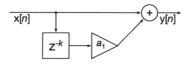

Figure 8.1

An example FIR filter, with a variable length (k) and gain (a) delay line.

8.1 FIR filters by way of example

Consider the FIR filter in Figure 8.1 with the difference equation $y[n]=x[n]+a_1x[n-k]$. For this example, let $k=1$, $a_1=0.5$ and the input $x_1[n]=[0.1, 0.4, 0.5, 0.2]$, which has a length of 4.

The first output is equal only to the input, since the delay tap will not generate an output. So, the output is given by $y_1[0]=x_1[0]+0.5\cdot x_1[0-1]=0.1+0.5\cdot0=0.1$. The next several outputs are:

$$y_1[1]=x_1[1]+0.5\cdot x_1[1-1]=0.4+0.5\cdot0.1=0.45 \tag{8.1a}$$

$$y_1[2]=x_1[2]+0.5\cdot x_1[2-1]=0.5+0.5\cdot0.4=0.7 \tag{8.1b}$$

$$y_1[3]=x_1[3]+0.5\cdot x_1[3-1]=0.2+0.5\cdot0.5=0.45 \tag{8.1c}$$

$$y_1[4]=x_1[4]+0.5\cdot x_1[4-1]=0.0+0.5\cdot0.2=0.1 \tag{8.1d}$$

Once the sixth sample is reached, the filter is flushed, and zero-valued inputs result in zero-valued outputs. Figure 8.2 (a) shows both the input and output of this filter. It can be seen in this case that the length of the output is equal to the length of $x_1[n]$ plus k. More generally, **the output of an FIR filter has a sequence length that is equal to the length of the input plus the order of the filter.**

Now consider a different input, $x_2[n]=[0.5, -0.4, 0.3, -0.2, 0.1]$, which has a length of five (we can, therefore, predict the length of $y_2[n]$ to be $5+1=6$ samples). The output, y_2, is as follows:

$$y_2[0]=x_2[0]+0.5\cdot x_2[0-1]=+0.5+0.5\cdot0.0=+0.5 \tag{8.2a}$$

$$y_2[1]=x_2[1]+0.5\cdot x_2[1-1]=-0.4+0.5\cdot0.5=-0.15 \tag{8.2b}$$

$$y_2[2]=x_2[2]+0.5\cdot x_2[2-1]=+0.3-0.5\cdot0.4=+0.1 \tag{8.2c}$$

$$y_2[3]=x_2[3]+0.5\cdot x_2[3-1]=-0.2+0.5\cdot0.3=-0.05 \tag{8.2d}$$

$$y_2[4] = x_2[4] + 0.5 \cdot x_2[4-1] = +0.1 - 0.5 \cdot 0.2 = 0.0 \tag{8.2e}$$

$$y_2[5] = x_2[5] + 0.5 \cdot x_2[5-1] = 0.0 + 0.5 \cdot 0.1 = 0.05 \tag{8.2f}$$

The output is the sequence $y_2 = [0.5, -0.15, 0.1, -0.05, 0.0, 0.05]$ and is shown (along with the input) in Figure 8.2(b). Of important note are the respective frequencies of x_1 and x_2. While these are not pure sinusoids and, therefore, comprise multiple frequencies, we can see that x_1 does not change polarity, while x_2 changes polarity every sample. From this simple observation, we can roughly characterize x_1 as LF and x_2 as HF. Next, turn your attention to the relative amplitudes of y_1 and y_2 compared to x_1 and x_2. It should be noted that y_1 has an increased amplitude (compared to x_1), while y_2 has an attenuated amplitude. In other words, this first-order FIR filter is boosting LFs while attenuating HFs.

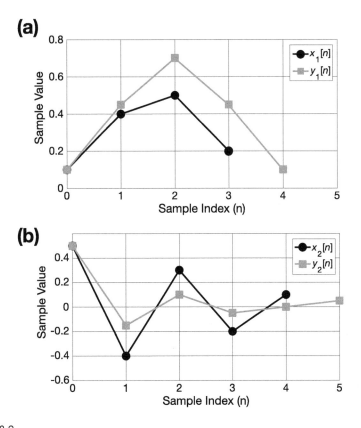

Figure 8.2

The FIR filter in Figure 8.1 results in (a) a boost of amplitude for an LF signal, and (b) an amplitude cut for an HF signal.

8.2 Impulse response

In the previous section, it was asserted that the output length is the sum of the filter order and the input length; therefore, it stands to reason that the IR of an FIR filter, $h[n]$, would have (finite) length $N+1$, where N is the filter order (and 1 is the length of $\delta[n]$). Using the same first-order LPF described in Chapter 8.1, the IR is easily calculated as

$$h[0] = \delta[0] + 0.5 \cdot \delta[0-1] = 1.0 + 0.5 \cdot 0.0 = 1.0 \tag{8.3a}$$

$$h[1] = \delta[1] + 0.5 \cdot \delta[1-1] = 0.0 + 0.5 \cdot 1.0 = 0.5 \tag{8.3b}$$

Importantly, the values of the IR match the gains of the taps of the filter, when sorted by increasing order. The gains corresponding to each filter tap is also known as the filter's *coefficients*. And more generally, **FIR filter coefficients are equal to the filter's IR, when each tap corresponds to a unique delay element, z^{-k}.** Under this condition, the coefficients are labeled with increasing subscripts, from 0, which represents the no-delay tap, up to the highest subscript, which is the order of the filter N. For the example above, we can write h in terms of the gains (where we assume a gain of $a_0 = 1$ for the top tap, since there is no gain explicitly diagramed): $h[n] = [1.0, 0.5] = [a_0, a_1]$.

8.3 Convolution

Given a filter's set of coefficients and some arbitrary input, it would be convenient to be able to compute the output algorithmically. Consider the generic third-order FIR filter in Figure 8.3. Since it was just shown that the IR is determined from the coefficients, we know the IR will be $h[n] = [a_0, a_1, a_2, a_3]$. Let's calculate the first several outputs for an arbitrary input, x.

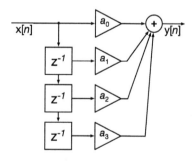

Figure 8.3

A third-order FIR filter.

$$y[0] = h[0] \cdot x[0] \qquad (8.4a)$$

$$y[1] = h[0] \cdot x[1] + h[1] \cdot x[0] \qquad (8.4b)$$

$$y[2] = h[0] \cdot x[2] + h[1] \cdot x[1] + h[2] \cdot x[0] \qquad (8.4c)$$

$$y[3] = h[0] \cdot x[3] + h[1] \cdot x[2] + h[2] \cdot x[1] + h[3] \cdot x[0] \qquad (8.4d)$$

Hopefully, you will see that a pattern has emerged. Let's just consider the last line, when $n = 3$. First, each term in the summation will be labeled with a k-value. Note that the k-value corresponds with the order of that tap. For example, the third-order tap with coefficient value a_3 corresponds to $h[3]$, which is part of the '$k = 3$' term.

$$\underbrace{y[3]}_{n=3} = \underbrace{h[0] \cdot x[3]}_{k=0} + \underbrace{h[1] \cdot x[2]}_{k=1} + \underbrace{h[2] \cdot x[1]}_{k=2} + \underbrace{h[3] \cdot x[0]}_{k=3} \qquad (8.5)$$

On further inspection, it can be seen that the argument of y matches with n, while the argument of h matches with k. The argument of x depends on both n and k, and is equal to $n-k$. The previous equation can be rewritten more generically as

$$y[n] = \underbrace{h[k] \cdot x[n-k]}_{k=0} + \underbrace{h[k] \cdot x[n-k]}_{k=1} + \underbrace{h[k] \cdot x[n-k]}_{k=2} + \underbrace{h[k] \cdot x[n-k]}_{k=3} \qquad (8.6)$$

Notice that the four terms are now identical! But while this equation is generalized for any-length input, it is fixed to a third-order filter. To make this equation even more generic, we should automatically sum up all of the terms, irrespective of the number of taps, using an accumulator. The IR, $h[n]$, houses the filter's coefficients from a_0 (at $h[0]$) to a_N (at $h[N]$).

$$y[n] = \sum_{k=-\infty}^{\infty} h[k] \cdot x[n-k] \overset{\text{def}}{=} h[n] * x[n] \qquad (8.7)$$

The process for computing the output of an FIR filter y for an input x with filter coefficients h is known as *convolution*. The convolution operator is denoted by '*'. Notice that inside the sum, the variable k is used to represent the sample index (instead of n). This change of variables is because the index of the output corresponds with the shift amount inside the summation, and it makes sense for y to be in terms of n.

$$y[n] = h[n] * x[n] = \sum_{k=-\infty}^{\infty} h \underbrace{[k]}_{\substack{\text{sample} \\ \text{index}}} \cdot x \begin{bmatrix} \underbrace{n}_{\text{shift}} \overset{\substack{\text{time} \\ \text{rev.}}}{-} k \end{bmatrix} \qquad (8.8)$$

For every shift amount, all of the overlapping samples of x and h are multiplied then their products summed, producing a single value to assign to y. As long as h and x are both real signals, then y is guaranteed to be real, as well.

8.3.1 Properties of convolution

Convolution is **commutative**, whereby $h[n]*x[n]=x[n]*h[n]$. Under the commutative property, the following is true (note how x and h are swapped):

$$h[n]*x[n]=\sum_{k=-\infty}^{\infty}x[k]\cdot h[n-k] \tag{8.9}$$

Convolution is also **associative**, whereby the order of operations does not matter.

$$(x[n]*h[n])*g[n]=x[n]*(h[n]*g[n]) \tag{8.10}$$

Finally, convolution is **linear**, so the operation can be arithmetically distributed or scaled, with no change to the output, whereby

$$x[n]*(a\cdot h[n]+b\cdot g[n])=a\cdot(x[n]*h[n])+b\cdot(x[n]*g[n]) \tag{8.11}$$

Convolution is representable in a block diagram, depicted as a box containing the IR of the system. In Figure 8.4, you can see each of the properties of convolution

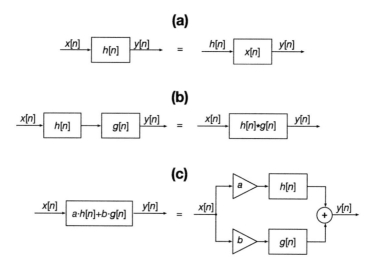

Figure 8.4

These block diagrams, containing the convolution process, depict the three properties of convolution, namely, (a) commutative property, (b) associative property, and (c) linear property. In all three cases, the block diagram on the left is equivalent to the block diagram on the right.

represented in block diagram form, (a) the commutative property, (b) the associative property, and (c) the linear property.

8.3.2 Example: convolving two signals

This example shows the convolution of two signals, $x = [0.1, 0.2, 0.0]$ and $h = [0.2, 0.0, -0.2]$, shown in Figure 8.5(a). During the convolution process x is kept fixed, while h is time-reversed, shown in Figure 8.5(b). The first output, $y[0]$, occurs when the last sample of the flipped signal overlaps with the first sample of the fixed signal, and is equal to their product, as shown in Figure 8.5(c). Each subsequent output corresponds with a sequential shift of the flipped signal. The value of each next output is a summation of the products of all overlapping samples, shown in Figure 8.5(d–f). The math for solving the convolution is as follows:

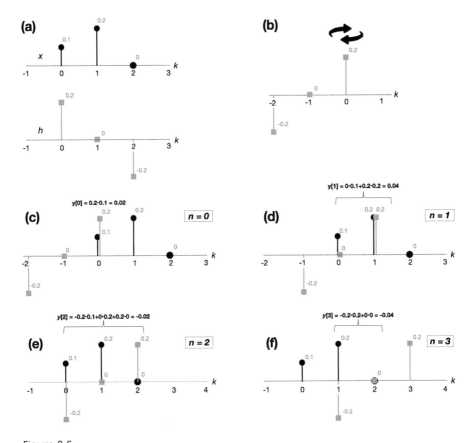

Figure 8.5

(a) Two signals are convolved, $x[n]$ (black) and $h[n]$ (gray), which is time-reversed, as shown in (b). In (c), only a single sample overlaps at $k = 0$.

$$y[0] = \sum_{k=0}^{0} x[k] \cdot h[0-k] = 0.1 \cdot 0.2 = 0.02 \qquad (8.12a)$$

$$y[1] = \sum_{k=0}^{1} x[k] \cdot h[1-k] = 0.1 \cdot 0.0 + 0.2 \cdot 0.2 = 0.04 \qquad (8.12b)$$

$$y[2] = \sum_{k=0}^{2} x[k] \cdot h[2-k] = 0.1 \cdot (-0.2) + 0.2 \cdot 0.0 + 0.0 \cdot 0.2 = -0.02 \qquad (8.12c)$$

$$y[3] = \sum_{k=1}^{3} x[k] \cdot h[3-k] = 0.2 \cdot (-0.2) + 0.0 \cdot 0.0 = -0.04 \qquad (8.12d)$$

$$y[4] = \ldots \qquad (8.12e)$$

One final shift amount is omitted from the math and figure; try to compute $y[4]$ on your own! Note the length of y (5 elements) is predictable based on the lengths of x and h (3 elements, each). In general, the number of elements in the output sequence is equal to the sums of the lengths of the two signals, minus one; in this example, $\text{len}(y) = \text{len}(x) + \text{len}(h) - 1 = 3 + 3 - 1 = 5$.

8.3.3 Programming example: vocal processing

Convolution is a fundamental process in audio and expresses how two signals interact, or how a signal interacts with a system. For example, a microphone is often selected based on its tonal character. When an audio recording is made with the microphone, the sounds it captures become convolved with IR of the microphone. In this example, you will record your own voice, and convolve it with an IR that can be downloaded from DigitalAudioTheory.com (or simply use an IR that you have at your disposal). This IR is of a guitar cabinet and gives a low-pass characteristic to the recording.

```
% Setup the audio recorder
Fs=44100;
nbits=16;
nchans=1;
r = audiorecorder(Fs, nbits, nchans);

% Record 2 sec of audio
recordblocking(r, 2);
% for Matlab use:
x = r.getaudiodata;
% for Octave use:
x = getaudiodata(r);

% Audition the recording
play(r);
```

8. FIR filters

```
% Load the ir
h = audioread('IR.wav');


% Prepare Output
len = length(x)+length(h)-1;
y=zeros(len,1);

% for every possible shift amount
for n=0:len-1
    % for every k from 0 to n; except before complete overlap
    % when we can stop at the length of h
    for k=0:min(length(h)-1,n)
  % check to see if h has shifted beyond the bounds of x
        if ((n-k)+1 < length(x))
% '+1' in every array argument since Matlab sequences start
% at index 1 (and not 0, like other programming languages)
            y(n+1) = y(n+1)+ h(k+1)*x((n-k)+1);
        end
    end
end


% Audition the output
soundsc(y, Fs)


% Built-In Command for convolution
y2=conv(x,h);
soundsc(y2, Fs)
```

8.4 Cross-correlation

There is a related function to convolution, known as *cross-correlation*, which is given by

$$x[n] \star h[n] = \sum_{k=-\infty}^{\infty} x[k] \cdot h[n+k] \qquad (8.13)$$

Notice that the only difference between correlation ($\star$) and convolution ($*$) is that the shifted signal (in this case, h) is **not** time-reversed. This simple change results in x and h being compared rather than combined. The cross-correlation operator results in a moving similarity score that compares the two signals at different *lags*, or the time-shift differences between two signals. This generates two important outcomes: (1) how much one signal resembles the other; and (2) when peak similarity occurs. The value of the cross-correlation is maximal when the two signals are time-aligned, resulting in the most amount of overlap. Polarity of the two signals does not change the magnitude of the output, only the sign of it. Two important uses of cross-correlation in digital audio are discussed below.

8.4.1 Programming example: time-delay estimation

It is not uncommon in digital audio signal processing to have multiple recordings of the same source that are not time-aligned. This could be due to microphone spacing or effects processing that introduces delays into a signal path. Cross-correlation provides a way to estimate the time-delay difference of two signals. Consider two sequences, $x[n]$ and $w[n]$ that are both recordings of a drum kit (downloadable from DigitalAudioTheory.com). The sequence w was recorded from 20 feet away to capture room effects but resulting in a delay of $t_d = \dfrac{20\ ft}{1{,}131\dfrac{ft}{s}} = 17.7\ ms$. If the microphone distance was not known to us beforehand and these were mixed together without a delay compensation, some frequencies within the two signals would be out-of-phase. We will estimate the time-delay using cross-correlation.

```
cx=audioread('drum.wav');
w=audioread('drum_delay.wav');

[similarity,lag] = xcorr(w,x);
[~,I] = max(abs(similarity));

plot(lag,similarity)
title(sprintf('Delay of %f sec',lag(I)/44100))
```

Making use of the built-in function **xcorr()** to quickly compute the cross-correlation, it returns the similarity value and corresponding lags. The lag corresponding with peak similarity is our estimated time delay, here, about 17.1 ms.

8.4.2 Programming example: matched filtering

Cross-correlation can also be used to perform *matched filtering*, or the process of identifying a pattern or template within a signal. This is common in audio for transient detection. In this example, you will use a template of a snare and cross-correlate that template with a drum-loop. Very large similarity scores indicate the presence of that snare, which will be plotted overlaid upon the drum-loop waveform.

```
x=audioread('drum.wav');
s=audioread('snare_hit.wav');

[similarity,lag] = xcorr(x,s);

I = find(similarity>50);

plot(x)
hold on;
stem(lag(I), ones(size(I)))
```

8.5 FIR filter phase

Thus far when we have come across frequency plots, we have only discussed them in terms of magnitude. However, we can also consider the phase of a signal or system.

While the magnitude indicates <u>how much</u> of a given frequency is passed, the phase indicates <u>how long</u> a frequency takes to pass. Specifically, the phase of a system signifies the number of cycles required by a particular frequency to pass through the system. Complicating matters is the fact that different frequencies cycle through phase at different rates – that is, much more time is needed for a 20 Hz signal to cycle through 2π rad than for a 10 kHz signal. However, if we instead look at the derivative of the phase, then we get a measure of a system's phase distortion, known as its *group delay*.

Group delay has units of time in a continuous system or units of samples in a digital system and indicates the transmission time at various frequencies. The group delay is calculated as the negative derivative of the phase; since phase is given in units of rad while frequency is given in rad/sec, then applying the derivative to phase leaves just sec

$$\overset{sec}{\tau_g} = -\frac{d\overset{rad}{\phi}}{d\underset{\frac{rad}{sec}}{\omega}} \tag{8.14}$$

All frequencies are delayed by some amount when a signal passes through a digital audio system, if that system is more complex than simple gain. When there is a variation in the amount of delay for different frequencies, that system is said to introduce *phase distortion*, which occurs when the group delay is not constant for all frequencies. Phase distortion changes the waveshape of a signal and alters the timbre of a transient or note attack. Phase distortion becomes notable in mid-range frequencies at 1 ms, but is not noticeable in LFs or HFs until 2–3 ms [1].

8.5.1 Linear phase

Under certain circumstances, the group delay for all frequencies is the same, which is called constant group delay. The phase of a constant group delay system can be obtained by integration, which will always result in a straight line. In this special case, the system exhibit *linear phase*, and it delays all frequencies by the same amount. Only an FIR filter can exhibit linear phase, and this only occurs when the coefficients (and IR) of the filter are symmetric:

$$h[n] = h[N-n-1], n = 0, 1, \ldots, N-1 \tag{8.15}$$

The group delay of a linear phase filter is simply a constant value, sometimes called its *latency*, t_d, which is given by:

$$t_d = \frac{N-1}{2 \cdot f_s} \tag{8.16}$$

For a proportionally longer filter (increased order, N, or decreased sample rate, f_S), the latency increases. Some latency can be a tolerable by-product of eliminating phase distortion, even in real-time systems. This is especially true if the latency is less than 10 to 15 ms, which is a buffer size around 512 samples for $f_S = 44.1$ kHz, meaning the largest number of taps that a linear phase FIR filter should have at this sampling rate is $N = 1,024$ taps.

However, one more pernicious artifact is introduced with symmetric (linear phase) filters. Since the coefficients are symmetric, the IR must also be symmetric, implying that any ringing that occurs after the mid-point of the IR must also occur prior to the mid-point, an artifact known as *pre-echo*. Pre-echo is a low-level ringing that occurs <u>prior to</u> the onset of a signal, which can be heard leading up to a transient (especially in a silent background) and has the effect of softening a sharp transient. More drastic EQ curves (e.g., steeper transitions, lower stopbands, flatter passbands) result in greater pre-echo, compared to more gradual curves. So, while linear phase filters are ideal at preserving the overall waveshape of the signal they affect, they may require a larger order, impose a larger delay, and introduce pre-echo when compared to other types of filters.

8.5.2 Minimum phase

An alternative to a linear phase filter is a *minimum phase* filter. If some phase distortion is tolerable, then delay and pre-echo can be eliminated. A minimum phase filter has the shallowest phase slope of any FIR filter, and therefore has the lowest group delay, albeit at varying amounts, depending on frequency. The coefficients (and IR) of a minimum phase filter resemble an acoustic IR, with an instantaneous output followed by ringing, and importantly, exhibiting minimal pre-echo. Finally, minimum phase filters and their inverse are both guaranteed to be stable, which is a unique property of minimum phase filters.

8.6 Designing FIR filters

There are several different FIR filter design methods, but fortunately, the digital signal processing toolboxes in software packages such as MATLAB® or Octave provide tools for easily implementing these different design methods. Designing a filter requires the specification of the filter *prototype*, or defining the desired filter in terms of the passband and stopband frequency regions, and their *ripple* and *rejection* in dB, respectively. The pass-band ripple is the maximum amount of tolerable variation (e.g., ±1 dB), while the stop-band rejection is the maximum amount of allowable attenuation (e.g., −80 dB). A smaller ripple, greater rejection, and smaller transition band all indicate a larger filter order, N. While finding the best filter order for a specific application can be a bit of trial and error, a good starting point is given by the Harris approximation [2]:

$$N \approx \frac{A_{reject} \cdot f_S}{22 \cdot BW_{transition}} \tag{8.17}$$

where A_{reject} is the amount of stop-band rejection (in dB), f_S is the sampling rate, and $BW_{transition}$ is the bandwidth of the transition band, or $\lfloor f_{pass}-f_{stop}\rfloor$. Other specifications may be required for certain design methods, but these are the typical parameters.

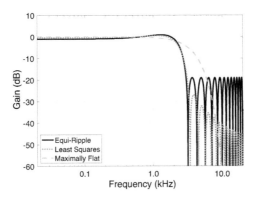

Figure 8.6

Different filter designs exhibit different properties, such as the amount of ripple (Maximally Flat has the lowest) to the amount of stopband rejection (here, Least Squares has the most).

To compare a few different FIR design methods, the Filter Designer in MATLAB® was used according to the following prototype: f_{pass} = 2.0 kHz, f_{stop} = 3.0 kHz, and f_S = 44.1 kHz. If we plug a transition band of 1.0 kHz and select a modest rejection ratio of 20 dB, then we get an order of N = 32. The Equi-ripple, Least Squares, and Maximally Flat designs are shown in Figure 8.6. If the order is specified (as it is here), then the ripple and rejection are not tunable, and vice versa.

8.6.1 Programming example: pre-echo

Two common types of FIR filters are linear phase and minimum phase filters. Linear phase filters have no phase distortion, while minimum phase filters have the lowest delay. In this example, an FIR filter will be designed using a filter design and analysis tool (**filterDesigner** in MATLAB®, which is unfortunately unavailable in Octave at the time of this writing). Since the filter effects are more easily heard with transients, a recording of castanets in silence will be used, available for download from DigitalAudioTheory.com.

Design the filter

First, open the filter designer and select the following options:
- Response Type: Lowpass
- Design Method: Least-squares
- Filter Order: 256
- Sample Rate: 44,100
- f_{pass}: 1,000
- f_{stop}: 2,000

Click 'Design' and you will see the frequency response of the filter. Note that this is a linear phase filter, so the phase response is a straight line (in the pass-band), the group delay is constant (128 samples), and the impulse response is symmetric. To use this filter in the workspace, click 'File > Export...' and change the name of these coefficients to a_lp (for linear phase). You will use these coefficients to also generate a minimum phase version of the same filter, using the command **firminphase()**. The castanets will be filtered with our two FIR filters with the command **filter()**, which takes as arguments the feed-forward coefficients, the feed-back coefficients (there are none, so pass '1'), and the signal to be filtered.

```
a_mp = firminphase(a_lp);
[x,fs] = audioread('castanets.wav');
sound(filter(a_lp,1,x),fs);
pause(6);
sound(filter(a_mp,1,x),fs);

grpdelay(a_mp,1,fs,fs); xlim([0 1000]);
```

Listen to the two outputs – can you hear a difference between linear phase and minimum phase? In this case, to best preserve the transient, we would likely select the minimum phase filter, which preserves the sharpness of the attack. However, examine the group delay of this filter – it is highly non-linear, even in the pass-band. The **grpdelay()** function takes as arguments the feed-forward and feed-backward coefficients, the frequency resolution and sampling rate, which are both f_S.

8.7 Challenges

For the following 5 problems, use the filter with IR:
$h[n] = [0, -0.02, -0.04, -0.08, 0.90, -0.08, -0.04, -0.02, 0]$.

1. Draw a feed-forward block diagram for this filter.
2. What is the order of this filter?
3. Is this filter linear phase (and why, or why not)?
4. A step is introduced to the input of this filter, $x[n] = [0, 0, 1.0, 1.0]$. Compute the output via convolution of x and h.
5. Draw or plot the step response found in Challenge #4.
6. Using a filter design tool, design a DC blocking filter at a sample rate of $f_S = 44.1$ kHz, with a delay of no more than 10 ms and a rejection of at least 99.9%
    a. What design method did you choose? Why?
    b. Is this filter linear phase?
    c. Plot the IR.
    d. What is the lowest frequency in the pass-band?
    e. What is the delay of your filter?
7. For the random sequence $r[n] = [-0.3, 0.1, 0.8, -0.5]$, compute the auto-correlation of this signal (Hint: auto-correlation is the cross-correlation of a sequence with itself).

8.8 Project – FIR filters

In this lab, you will implement a first-order FIR filter that will have the form

$$y[n] = a_0 x[n] + a_1 x[n-1]$$

Implementing such a filter in code would normally involve a for loop to iterate over all of the available input samples, x. For example, in a DAW it is typical to receive some number of input samples that are selected by the user (e.g., 128, 256, 512, etc.). Importantly, all of the filter states (the values held in the delays) need to be preserved not only within the scope of the for-loop but outside it too so that the filter is continuous between blocks of audio arriving from the DAW.

```
xnm1=0; % zero out filter state
for (n=1:length(x)+1)  % +1 to account for filter order
    y(n) = a0*x(n) + a1*xnm1;
    xnm1=x(n);
end
```

However, sample-by-sample operations are rather slow in MATLAB® or Octave. These types of computing environments are optimized to perform efficient array and matrix operations, so we will take advantage of this.

```
y = a0 * x + a1 * xnm1;
```

Here, y, x, and *xnm1* are all vectors. To create *xnm1*, we will zero-pad it with a single zero, and then also add a zero to the end of x to make it the same length.

```
xnm1 = [0, x];
x = [x, 0];
```

In this project, you will implement a first-order FIR filter using varying coefficients a0 and a1.

Assignment

1. Implement a first-order FIR filter using coefficients a0 = 0.5 and a1 = 0.5
2. Generate a noise signal using **rand()** that is 2 seconds in length. Be sure to scale and shift to put the random signal in the range of ±1.
3. Filter the noise signal with your FIR filter. Plot the input and output signals in a 2 × 1 subplot. Play the two signals and take note of their differences. Describe your observations – what type of filter is this?
4. Repeat step 3 with an audio file that you load in. How does the EQ change when inserted?
5. Repeat steps 1 through 4 with coefficients a0 = 0.5 and a1 = −0.5
   In the examples above, we only worked with a first-order filter. For the next part, you will implement a second-order FIR filter using coefficients of your choosing.

6. Repeat steps 1–3 using your second-order FIR filter coefficients. You will need to add a new vector, xnm2, and change how xnm1 and x are defined.
   a. Does this filter sound how you expected? How is it similar or different to the first-order filter?

Bibliography

[1]   Blauert J, Laws P. "Group Delay Distortions in Electroacoustical Systems." *Journal of the Acoustical Society of America* Vol. 63, No. 5 (1978): 1478–1483.
[2]   Harris F. *Multirate Signal Processing for Communication Systems*, Prentice Hill, Upper Saddle River, NJ, 2004, p. 216.

9

z-Domain

Up until this point, we have not had the tools to analyze a filter's frequency response – for example, in the project for Chapter 8, the impact of the filter was determined by listening for the effects of different coefficients. However, in this chapter, we will learn how to predict a filter's frequency response based on the block diagram and difference equation. In particular, a filter's *magnitude response* (the impact of a filter on the level at different frequencies) and *phase response* (the filter's impact on the timing of different frequencies) will be derived from the frequency response.

If you have had previous experience with a Laplace transformation using the complex *s* variable, then this chapter should be intuitive. We will introduce a complex *z* variable that behaves in the same manner, except for digital systems rather than continuous ones. Importantly, by virtue of being in a digital system, we will not be using the *s*-plane, but rather the *z*-plane (on which sits the unit circle). If you have no experience with Laplace transformations, no worries – this chapter provides a comprehensive theoretical basis for frequency domain analysis of digital filters.

9.1 Frequency response

Up until this point, we've discussed the frequency response of a filter in terms of its generic shape – for example, high-pass, low-pass, etc. When characterizing the relative level across different frequencies of a filter, that aspect of the frequency response is the *magnitude response*. The magnitude response gives the exact impact that a filter has on the amplitude of any given frequency. Think back to chapter 2 when phasors were discussed – a complex phasor could be separated into its magnitude and phase portions. Well, frequency response behaves in the same manner. Truly, a filter's frequency response is complex, and so in addition to having a magnitude portion, it also has a phase portion. The *phase response* of a filter is related to the amount of delay experienced by each frequency. The magnitude and phase responses can be combined into the complex frequency response (and vice versa). To analyze a filter, we first derive its frequency response, then take the absolute value to extract the magnitude response or the angle (using the arc-tangent) to extract the phase response. Consider the simple FIR filter, given as

$$y[n] = x[n] + a_k x[n-k] \tag{9.1}$$

where $x[n]$ is a sinusoid of arbitrary frequency, f. This filter has an IR of $h[n] = [1, 0, ..., 0, a_k]$ and an order of k. Even though an audio signal, x will always necessarily be real, let's consider it as a complex phasor. This will be assumed only to make the math wieldier – this process can be re-computed with real signals, to the same conclusion. The signal $x[n]$ will be written in exponential form, as

$$y[n] = e^{\overbrace{j2\pi f n \over f_s}^{x[n]}} + a_k\, e^{\overbrace{j2\pi f(n-k) \over f_s}^{x[n-k]}} \tag{9.2a}$$

$$y[n] = e^{j2\pi f n \over f_s} + a_k e^{-{j2\pi f k \over f_s}} e^{j2\pi f n \over f_s} \tag{9.2b}$$

$$y[n] = \underbrace{e^{j2\pi f n \over f_s}}_{x[n]}\left[1 + a_k e^{-{j2\pi f k \over f_s}}\right] \tag{9.2c}$$

It can be seen in the last line that a common factor can be pulled out, and this factor is equivalent to $x[n]$. Interestingly, we've rewritten the difference equation as y equals x times the term contained in the square brackets, which includes the filter coefficients a_0 and a_k. This is interesting because it lets us separate the input, $x[n]$, leaving only a portion that is entirely "agnostic" to the input. This portion (the part in brackets) will be called $H(f)$, or frequently $H(\omega)$, which can be expressed as

$$H(f) = 1 + a_k e^{-{j2\pi f k \over f_s}} \tag{9.3a}$$

or

$$H(\omega) = 1 + a_k e^{-j\omega k} \tag{9.3b}$$

where

$$\omega = \frac{2\pi f}{f_s} \tag{9.3c}$$

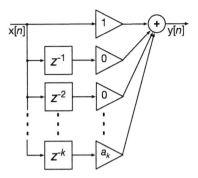

Figure 9.1

The block diagram realization of a filter with frequency response $H(\omega)=1+a_k e^{-j\omega k}$.

Pop-quiz: Is H (which contains an exponential term) a phasor – why or why not?
Answer: In fact, H is not a phasor because the phase is not changing – it is the constant $-j2\pi fk/f_S$, where k is the amount of delay. Interestingly, an FIR block diagram is directly implementable from reading the frequency response, which says that the pass-through tap (no delay) has a gain of 1 while the k-delay tap has a gain of a_k.

Additionally, the IR, $h[n]$, can be determined by inspection from H – at the no-delay position of the IR, $x[0]$, the gain is 1, and at the k-delay position of the IR, the gain is a_k. In between the no-delay and k-delay positions, the IR is filled with zeroes, since each of those could be imagined as taps with gains of 0, as shown in Figure 9.1. Note the congruency between h and H; these are the same letter (just with different case) for a reason – H is, in fact, the frequency response of h. In general, lower case letters will be used to represent sample-based sequences (for example, $h[n]$) while upper case letters will be used to represent frequency responses (for example, $H(f)$).

9.2 Magnitude response

When dealing with complex phasors, the magnitude of the phasor was determined by taking the square root of the sum of the squares of the real and imaginary parts. This same principle applies when determining the magnitude of frequency response. It is an easier proof if we first convert to rectangular form:

$$\left|H(\omega)\right|=\left|1+a_k e^{-j\omega k}\right|=\left|1+a_k\cos(\omega k)-ja_k\sin(\omega k)\right| \tag{9.4}$$

In this form, we can easily extract the real and imaginary parts:

$$\text{Re}\{H\}=1+a_k\cos(\omega k) \tag{9.5a}$$

$$\text{Im}\{H\}=-a_k\sin(\omega k) \tag{9.5b}$$

Then, to determine the magnitude response, we square each of these parts, sum them, and take the square root:

$$|H(\omega)| = \sqrt{\text{Re}\{H\}^2 + \text{Im}\{H\}^2} \qquad (9.6a)$$

$$|H(\omega)| = \sqrt{(1 + a_k \cos(\omega k))^2 + (-a_k \sin(\omega k))^2} \qquad (9.6b)$$

$$|H(\omega)| = \sqrt{1 + 2a_k \cos(\omega k) + a_k^2 \cos^2(\omega k) + a_k^2 \sin^2(\omega k)} \qquad (9.6c)$$

$$|H(\omega)| = \sqrt{1 + 2a_k \cos(\omega k) + a_k^2} \qquad (9.6d)$$

At this point, we can substitute $2\pi f/f_S$ in place of ω:

$$|H(f)| = \sqrt{1 + 2a_k \cos\left(\frac{2\pi f k}{f_S}\right) + a_k^2} \qquad (9.7)$$

If we let $k = 1$, then the argument of the cosine ranges from 0 to π, meaning the cosine ranges from 1 to −1. If we let $a_k = 1$, then the magnitude ranges from 2 down to 0.

$$|H(f)| = \sqrt{2 + \underbrace{2\cos\left(\frac{2\pi f}{f_S}\right)}_{[-2,+2]}} \qquad (9.8)$$

With these specific parameters, then when f is close to 0, it has a magnitude of 2 (or +6 dB), and when it is near the Nyquist frequency, it has a magnitude near 0 (or −∞ dB). In other words, this specific case when $k = 1$ makes a low-boost, high-cut filter!

9.3 Comb filters

This form of filter is known as a *comb filter*. It is so called because the magnitude response, comprising one or more notches, resembles the shape of a comb, in which frequencies in fixed intervals can be selectively removed. A comb filter is the basis of a flanger effect, in which the comb frequency spacing is slowly modulated. This is achieved by changing the amount of delay, k, in the delay tap. In the following code, we will examine the impact on different k and a_k values on the magnitude response of the comb filter

9.3.1 Programming example: comb filters

First, let's plot the comb filter when $a_k = 1$ and $k = 1$. We already determined that this filter should provide some boost near 0 Hz and a big cut near f_N. We will select $f_S = 44.1$ kHz, so $f_N = 22.05$ kHz. In order to plot the magnitude response, which takes frequency, f, as a variable, we will generate a frequency array with spacing of 1 Hz.

```
ak=1.0;        % gain on the delay tap
k=1;   % amount of delay on the delay tap


Fs=48000;      % sample rate = 44.1 khz
f=[0:23999];% frequency array from DC to Nyquist

% magnitude response
Hmag=sqrt(1+2*ak*cos(2*pi*f*k/Fs)+ak^2);
% plotting and labeling axes
plot(f,20*log10(Hmag))
box on; grid on;
xlabel('Freq (Hz)');
ylabel('Mag (dB)');
title('Comb Filter');
xlim([f(1) f(end)])
```

The amount of LF boost could be lowered by applying a gain of 0.5 to the output. For this reason, this comb filter (when $k = 1$) is actually an LPF. What would happen if we sweep the k variable?

```
ak=1.0;        % gain on the delay tap
k=1;   % amount of delay on the delay tap


Fs=48000;      % sample rate = 44.1 khz
f=[0:23999];% frequency array from DC to Nyquist

% magnitude response
Hmag=sqrt(1+2*ak*cos(2*pi*f*k/Fs)+ak^2);

% plotting and labeling axes
plot(f,20*log10(Hmag))
box on; grid on;
xlabel('Freq (Hz)');
ylabel('Mag (dB)');
title('Comb Filter');
xlim([f(1) f(end)])
```

As shown in Figure 9.2, every increasing k value introduces a new peak or notch, with even k having no cut at the Nyquist frequency, while odd k values do have a cut at Nyquist. The spacing between these peaks and notches is also determined by k, given by f_N/k.

We can also examine the impact on sweeping the gain at the kth tap, by creating an array of a_k values, varying from -1 to 1. A delay of $k = 4$ will be used to make the effect of a_k obvious.

```
k=4;
ak=[-1:0.5:1];

for ak=-1:0.5:1
    Hmag=sqrt(1+2*ak*cos(2*pi*f*k/Fs)+ak^2);
    hold on;
    plot(f, 20*log10(Hmag));
end
```

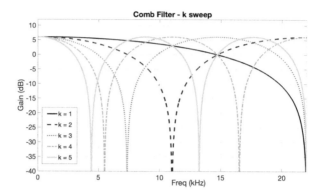

Figure 9.2

A comb filter is shown with $a_k = 1$ and k varying from 1 to 5. As k increases, the number of peaks and notches also increases.

```
title('Comb Filter - a sweep')
ylabel('Gain (dB)');
xlabel('Freq (Hz)');
box on; grid on;
axis([f(1) f(end) -40 12])
```

What happens to the magnitude response when a_k is negative? And what is the difference between a_k with low value compared to one with a higher value?

9.3.2 Programming example: removing ground hum

Ground hum can be introduced into an audio signal by having separate ground nodes within a circuit that actually contain a potential between them, or by electromagnetic interference that arises from electrical equipment or lighting. In either case, the symptom is the same – 120 Hz buzz with strong integer harmonics. This type of noise can be addressed with a comb filter, which targets frequencies with specific intervals. Let's assume a sample rate of $f_S = 48$ kHz, and the value $a_k = 1.0$ (maximum attenuation). The digital frequency location is given by

$$\omega = \frac{2\pi f}{f_S} = \frac{2\pi 120}{48,000} = 0.005\pi \text{ rad}$$

We want to place a notch at this frequency location, whereby $|H(\omega)| = 0$. Substituting in this value, $a_k = 1$, and $\omega = 0.005\pi$ and solving for k, we find,

$$0 = \sqrt{1 + 2\cos(0.005\pi k) + 1}$$

$$-1 = \cos(0.005\pi k)$$

$$\pi = 0.005\pi k$$

$$k = 200$$

Given this, can you determine what the IR will be?

$$h[n] = \left[1, \underbrace{0, \ldots, 0,}_{199} 1 \right]$$

The gain on the no-delay tap is 1, the gain on the 200th tap is $a_k = 1$, and every other tap in between has a gain of 0. In Chapter 9.1 we saw that the frequency response can be determined by visual inspection (with FIR filters) from the difference equation, which is,

$$H(\omega) = \underbrace{1}_{a_0} \cdot \underbrace{e^{0 \cdot j\omega}}_{tap\ 0} + \underbrace{1}_{a_{200}} \cdot \underbrace{e^{-200 \cdot j\omega}}_{tap\ 200}$$

Typically, on the zero-delay tap, the exponential would not be included since $e^0 = 1$, but it is shown here to help understand the equation. Below, this ground hum cancelling filter is implemented in code.

```
ak=1;
k=200;
Hmag=sqrt(1+2*ak*cos(2*pi*f*k/Fs)+ak^2);
plot(f, 20*log10(Hmag));
box on; grid on;
ylabel('Gain (dB)');
xlabel('Freq (Hz)');
title('Ground Hum Cancellation')
axis([f(1) f(end) -30 10])
```

9.4 z-Transform

In Chapter 7.4, the operator, z^{-k}, was introduced within a block diagram to represent a delay amount of k samples. Later in Chapter 7.4.2, we saw that delaying a signal x by k samples is equivalent to multiplying x by $e^{-j\omega k}$. Interestingly, we can think of the z^{-k} as either a delay operator or a complex variable, because it is both of these things! This duality of z is a powerful concept that will help transform between block diagrams and frequency response. More formally, z is defined in a digital system as,

$$z = A \cdot e^{j\omega T_S} = A \cdot e^{\frac{j2\pi f}{fs}} \tag{9.9}$$

While $e^{j\omega T_S}$ traces out just the unit circle (since it has a radius of 1), z on the other hand, is a variable that represents the entire complex plane, which is often called the z-plane for this very reason.

The z-transform is simply another way of looking at a filter or an input sequence. Since both sequences and filters can be transformed into the z-domain, we have another way to analyze a digital signal or system, in addition to a difference equation and block diagram. The definition of the z-transform for a causal digital sequence is,

$$X(z) = \sum_{n=0}^{N-1} x[n] \cdot z^{-n}$$
(9.10)

In other words, each subsequent sample of sequence $x[n]$ is multiplied by z^{-n}, which you will recall is simply a delay of n samples. For example, for the sequence $x[n]$ = [0.1, 0.2, 0.3, 0.4], the z-transform is given by

$$X(z) = 0.1z^{-0} + 0.2z^{-1} + 0.3z^{-2} + 0.4z^{-3}$$
(9.11)

Quite simply, this can be interpreted as, "value 0.1 is delayed by 0 samples, value 0.2 is delayed by 1 sample..." and so on. Furthermore, if we want to delay the entire sequence of x by three samples, this would be represented as $z^{-3}X(z)$, yielding the sequence: [0.0, 0.0, 0.0, 0.1, 0.2, 0.3, 0.4]. Note that similar to how H represents the frequency domain transformation of h here, we are also capitalizing X to indicate that it is no longer a sample domain representation, but rather a frequency domain representation (as the definition of z contains f and not n).

The z-transform of a digital system or filter gives us its transfer function. A *transfer function* of a filter is similar to its frequency response, in that a transfer function describes the effect that a filter has on an input sequence. Mathematically, this can be stated as a ratio of the output over the input:

$$H = \frac{\overset{output}{\overbrace{Y}}}{\underset{input}{\underbrace{X}}}$$
(9.12)

In Chapter 9.1 we saw that the frequency response, denoted by H, was a function of frequency, f. Unlike the difference equation, in which the filter delays and input sequence are intertwined, when represented as a frequency response or transfer function, the input is disassociated from the filter, allowing each to be characterized separately. But while the frequency response is in terms of $e^{j2\pi fT_S}$, the transfer function is in terms of z, meaning that it describes the filter anywhere on the z-plane, unlike the frequency response, which describes a filter only on the unit circle (this distinction will be revisited in Chapter 9.5).

Use of the z-transform makes solving the transfer function easy! Consider the comb filter from Figure 9.1, with the following difference equation: $y[n] = x[n] + a_k x[n-k]$. In order to determine the transfer function, we will transform the difference equation to the z-domain:

$$Y(z) = X(z) + a_k X(z) z^{-k}$$
(9.13a)

$$Y(z) = X(z) \cdot \left(1 + a_k z^{-k}\right)$$
(9.13b)

Just as in the case of frequency response, we see that Y equals X times *something*. Here that *something* is the transfer function, $H(z)$:

$$H(z) = 1 + a_k z^{-k} \tag{9.14}$$

If we consider z at a radius of $r = 1$ (on the unit circle) then we obtain the frequency response,

$$H(f) = H(z)\big|_{z=e^{j2\pi f/fs}} = 1 + a_k e^{-kj2\pi f/fs} \tag{9.15}$$

Hopefully, some patterns should be emerging. Previously, we saw that each sample of an FIR filter's IR corresponds directly to the gain of a tap with a corresponding delay. Here, it can be seen that the transfer function (for an FIR filter) also derives directly from the IR. At the end of the chapter, you will be challenged to convert between block diagram $\Leftrightarrow$ IR $\Leftrightarrow$ difference equation $\Leftrightarrow$ z-transform $\Leftrightarrow$ frequency response.

9.4.1 Properties of z-Transform

For the digital sequence $x[n]$, the z-Transform, $X(z)$, has the following properties.
  A **time-shift** of x by k samples is equivalent to multiplying X by z^k:

$$x[n-k] \overset{z}{\Leftrightarrow} z^{-k} \cdot X(z) \tag{9.16}$$

Convolution of x with h, whose z-Transform is $H(z)$, in the sample domain is equal to multiplication of X and H in the z-domain:

$$x[n] * h[n] \overset{z}{\Leftrightarrow} X(z) \cdot H(z) \tag{9.17}$$

Multiplication in the z-domain is **commutative**, whereby

$$H(z) \cdot X(z) = X(z) \cdot H(z) \tag{9.18}$$

Transfer functions in the z-domain are **associative**, whereby the order of operations does not matter. For example, when the z-Transform of $g[n]$ is $G(z)$:

$$\big(X(z) \cdot H(z)\big) \cdot G(z) = X(z) \cdot \big(H(z) \cdot G(z)\big) \tag{9.19}$$

Finally, transfer functions in the z-domain are **linear**, meaning they are scalable, additive and distributive:

$$X(z) \cdot \big(a \cdot H(z) + b \cdot G(z)\big) = a \cdot X(z) \cdot H(z) + b \cdot X(z) \cdot G(z) \tag{9.20}$$

9.4.2 Example: analyzing a cascade of filters

Consider the cascade of filters in Figure 9.3. The first filter $F(z)$ has IR $f[n]$ and the second filter $G(z)$ has IR $g[n]$. The input sequence is $x[n]$ with z-Transform $X(z)$. The intermediate output of F is given by,

$$w[n]=x[n]*f[n]\overset{z}{\Leftrightarrow}W(z)=X(z)\cdot F(z)$$

The output of the filter cascade is given by,

$$y[n]=w[n]*g[n]\overset{z}{\Leftrightarrow}Y(z)=W(z)\cdot G(z)$$

Problem: For the filters F and G defined as,

$$F(z)=a_0+a_1z^{-1}$$

$$G(z)=b_0+b_1z^{-1}$$

Determine the following:
- the transfer function $H(z)$,
- the difference equation $y[n]$ in terms of a_0, a_1, b_0, b_1, and x,
- the IR, h, of the filter cascade,
- the frequency response, $H(\omega)$,
- and the block diagram using two delays and three gains,

Solution: First, we will determine the transfer function H in terms of a_0, a_1, b_0, and b_1. This can be done by simply substituting in the equation for the intermediate output W into the equation for Y:

$$Y(z)=W(z)\cdot G(z)$$

$$Y(z)=X(z)\cdot F(z)\cdot G(z)$$

$$H(z)=\frac{Y(z)}{X(z)}=F(z)\cdot G(z)$$

Next, we will substitute in the definitions of F and G, relying on the linear property of transfer functions.

Figure 9.3

A cascade of two filters, F and G, for input x produces an output y with intermediate output w.

$$H(z) = F(z) \cdot G(z) = \left(a_0 + a_1 z^{-1}\right) \cdot \left(b_0 + b_1 z^{-1}\right)$$

$$H(z) = a_0 b_0 + \left(a_0 b_1 + a_1 b_0\right) z^{-1} + a_1 b_1 z^{-2}$$

Notice that while F and G are each first-order filters that, when they are in cascade, form a second-order filter.

The difference equation is found by multiplying through by $X(z)$ and then converting from the z-domain to sample domain, a process known as the *inverse z-Transform*.

$$Y(z) = H(z) \cdot X(z)$$

$$Y(z) = a_0 b_0 X(z) + \left(a_0 b_1 + a_1 b_0\right) z^{-1} X(z) + a_1 b_1 z^{-2} X(z)$$

Recall from the time-shift property, that $z^{-k}X(z)$ is equivalent in the sample domain to $x[n-k]$, so taking the inverse z-Transform, we obtain:

$$y[n] = a_0 b_0 x[n] + \left(a_0 b_1 + a_1 b_0\right) x[n-1] + a_1 b_1 x[n-2]$$

Then the IR can be determined through inspection by reading the gain coefficients for each tap of the filter:

$$h[n] = \left[a_0 b_0, \left(a_0 b_1 + a_1 b_0\right), a_1 b_1\right]$$

Next, we will determine the frequency response by analyzing on the unit circle, using the following substitution:

$$H(f) = H(z)\big|_{z=e^{j2\pi f/fs}}$$

Converting to the z-domain is as simple as applying z^{-k} to each subsequent term of the IR. However, it is common for the frequency response to be expressed with positive orders, so the transfer function will first be multiplied by z^2/z^2 before applying the substitution,

$$H(z) = a_0 b_0 + \left(a_0 b_1 + a_1 b_0\right) z^{-1} + a_1 b_1 z^{-2}$$

$$H(z) = \frac{z^2}{z^2} \cdot \left(a_0 b_0 + \left(a_0 b_1 + a_1 b_0\right) z^{-1} + a_1 b_1 z^{-2}\right)$$

$$H(z) = \frac{a_0 b_0 z^2 + \left(a_0 b_1 + a_1 b_0\right) z + a_1 b_1}{z^2}$$

$$H(f) = \frac{a_0 b_0 e^{2j2\pi f/fs} + \left(a_0 b_1 + a_1 b_0\right) e^{j2\pi f/fs} + a_1 b_1}{e^{2j2\pi f/fs}}$$

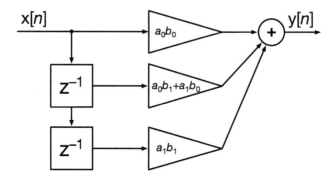

Figure 9.4

The block diagram realization for the filter $H(z) = F(z)G(z)$.

Last, the block diagram contains an accumulator with 3 taps, with 1 gain each. And since the order of this filter is 2, the minimum number of delays that can be used is also 2. The block diagram for this filter can be drawn as given in Figure 9.4.

9.5 Pole/zero plots

So, what is special about analyzing a filter right on the unit circle, as opposed to elsewhere on the z-plane? On the unit circle is the only place on the z-plane where a phasor is *critically stable*. Inside the unit circle, a phasor will decay to zero (which is still considered stable), while outside the unit circle, a phasor will "blow up" and become unstable, according to equation (1.10) from Chapter 1. To understand the reasoning behind this, let's start with the definition of z, but set it in motion by multiplying by n in its exponent.

$$z^n = \left(A \cdot e^{j\omega}\right)^n = A^n \cdot e^{j\omega n} \tag{9.21}$$

In the case that A is less than 1, then the amplitude will decay, and if A is greater than 1, then it will grow exponentially, as shown in Figure 9.5. Therefore, to analyze the frequency response of a filter, we analyze exactly at exactly a radius of 1.

Up to this point, we've used the z-plane to represent digital sequences, normally in the form of a sinusoid, but interestingly we can represent digital systems (such as a filter) on the z-plane as well! The roots of the transfer function, known as *poles* and *zeros*, provide salient information about both the magnitude and phase response, which can be plotted on the z-plane. This type of visualization is known as a pole/zero (or p/z) plot.

Zeros are roots of the numerator of the transfer function and are represented by 'o' on a p/z plot. Zeros have the effect of attenuating any frequencies near them. Since feed-forward taps always appear in the numerator of a transfer function, FIR filters will always have p/z plots with only zeros. A zero can appear in the z-plane either inside, on, or outside the unit circle and in all cases will be stable. In

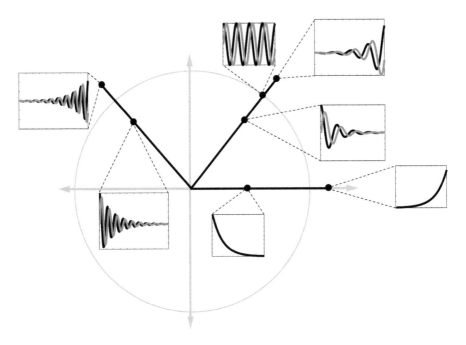

Figure 9.5

On the Real axis, the sine (Imaginary – gray) portion has no value, and when $A < 1$, the static cosine (Real – black) portion decays and grows when $A > 1$. Moving counter-clockwise off of the Real axis towards π, the frequency increases. Similarly, if the radius $A < 1$, then the phasor decays, and if the radius is $A > 1$, it grows.

fact, an FIR filter can never become unstable, since by definition its IR eventually reaches 0. When all of the zeros are inside the unit circle, the filter is said to exhibit minimum phase, or short latency. Minimum phase FIR filters are discussed in Chapter 8.5.2.

Poles are the roots of the denominator of a transfer function, and are represented by '×' on a p/z plot. Poles have the effect of boosting any frequencies near them. Any feed-back taps in a filter will appear in the denominator, and will therefore produce a pole. In Chapter 10, filters with feed-back taps will be discussed in detail, and transfer functions and p/z plots will be revisited at that point. Unlike zeros, poles can only appear <u>strictly</u> within the unit circle for the filter to be stable. A pole directly on the unit circle will result in a critically stable oscillation that does not grow exponentially, but also never decays to zero and is, therefore, also unstable, by definition.

9.5.1 Example: first-order LPF and HPF

A difference equation for a first-order FIR filter is given by

$$y[n] = x[n] - a_1 x[n-1]$$

Problem: Determine the following:
- the transfer function for the FIR filter,
- the magnitude response,
- the zero locations,
- draw the p/z plot on the z-plane.

Solution: First, the transfer function is computed by taking the z-transform, then solving for H:

$$Y(z) = X(z) - a_1 X(z) z^{-1}$$

$$Y(z) = X(z) \cdot \left(1 - a_1 z^{-1}\right)$$

$$H(z) = \frac{Y(z)}{X(z)} = 1 - a_1 z^{-1}$$

To solve for the magnitude response, we should apply the absolute value, then substitute $z = e^{j 2\pi f / fs}$. The magnitude response will make more sense with positive exponents, so we first multiply by z/z:

$$H(z) = \frac{z - a_1}{z}$$

$$|H(z)| = \frac{|z - a_1|}{|z|}$$

$$|H(f)| = |H(z)|_{z = e^{\frac{j2\pi f}{fs}}} = \frac{\left|e^{\frac{j2\pi f}{fs}} - a_1\right|}{\left|e^{\frac{j2\pi f}{fs}}\right|} = \left|e^{\frac{j2\pi f}{fs}} - a_1\right| = \left|e^{\frac{j2\pi f}{fs}} - a_1\right|$$

Note that, irrespective of frequency, the magnitude of the denominator will always equal 1; furthermore, for FIR/feed-forward filters, this will <u>always</u> be the case.

There are no roots of the denominator of the transfer function, except trivially at the origin. But it can be a good habit to mark poles here, simply to keep track of them, since there should be the same number of poles as zeros, and there should be the same number of both of these as the order of the filter. In this example, the order is 1, so we should anticipate 1 pole and 1 zero. The pole will be placed at the origin, $z_x = 0$ (we will use z_x to represent pole locations on the z-plane).

The zero location is at the root of the numerator of the transfer function. We can determine the root by setting the numerator to 0 and solving for z:

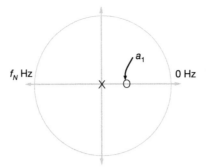

Figure 9.6

A p/z plot of the filter $y[n] = x[n] - a_1 x[n-1]$. Since this is an FIR filter, the pole is at the origin (no effect). The zero is located on the horizontal axis, at a value equal to the gain coefficient a_1.

$$z_o - a_1 = 0$$

$$z_o = a_1$$

Note the use of the subscript z_o to represent the zero locations. Since z is a complex variable, we have a clue as to the location on the z-plane. If a complex variable equals a real scalar, then the imaginary portion must be 0. This only occurs at $f = 0$ Hz or $f = f_N$ Hz; therefore, in this example, z_o is located on the horizontal axis, at a value of a_1, as shown in Figure 9-6.

Recall, that frequencies near a zero are attenuated, and with a_1 on the horizontal axis, the frequencies that will be affected most will be 0 Hz (if $a_1 > 0$), or f_N Hz (if $a_1 < 0$). In other words, depending on the sign of a_1, this filter can be configured as either an HPF or an LPF. Take a look again at the equation for the magnitude response:

$$|H(f)| = \left| e^{\frac{j2\pi f}{f_S}} - a_1 \right|$$

Quite literally, this tells us that the magnitude at location f is the difference (or distance) between that frequency location on the unit circle and the z_o at a_1. For example, if $f = 100$ Hz, and $f_S = 44,100$ Hz, and we select a value of $a_1 = 0.5$ (HPF), then the magnitude at this frequency is,

$$|H(100)| = \left| e^{\frac{j2\pi 100}{44,100}} - 0.5 \right| = 0.5001 = -6.019 \text{ dB}$$

Any frequency and sample rate can be plugged into this equation to determine the magnitude response. Or if this filter was configured as an LPF and a_1 was moved to the negative side of the horizontal axis at $a_1 = -0.5$, then the magnitude at 100 Hz would be

$$|H(100)| = \left| e^{\frac{j2\pi 100}{44,100}} - (-0.5) \right| = 1.5 = +3.522 \text{ dB}$$

9.5.2 Programming example: notch filter

For the next example, we will move the zero off of the horizontal axis, and to another frequency region, which will generate a notch corresponding to the frequency that is nearest to the zero. Consider the second-order FIR filter with the following IR:

$$h[n] = [1, \ -1, \ 1]$$

Problem: Determine the following:
- the difference equation,
- the transfer function,
- draw the p/z plot on the z-plane,
- plot the magnitude response in MATLAB® or Octave.

Solution: Since this is an FIR filter, the coefficients for each delay can be read directly from the IR:

$$y[n] = x[n] - x[n-1] + x[n-2]$$

To calculate the transfer function, first, we must take the z-Transform, then divide the output by the input. Then, we will multiply by z^2/z^2 to make the exponentials positive:

$$Y(z) = X(z) - X(z)z^{-1} + X(z)z^{-2}$$

$$Y(z) = X(z) \cdot \left(1 - z^{-1} + z^{-2}\right)$$

$$H(z) = \frac{Y(z)}{X(z)} = 1 - z^{-1} + z^{-2}$$

$$H(z) = \frac{\overbrace{z^2 - z + 1}^{\text{roots are } z_o \text{ locations}}}{\underbrace{z^2}_{\text{2 poles at origin}}}$$

Next, to draw the p/z plot, we need to determine the roots of the numerator (zeros) and denominator (poles). The poles are easy – there are two at the origin. To find the roots of the numerator, we set the numerator equal to zero and solve for z:

$$0 = z_0^2 - z_0 + 1$$

This is a little bit tricky to FOIL in reverse. The first term of each factor must be z, clearly. But to get a negative middle term, the second term of each factor must be complex. For two complex numbers to multiply to 1, their exponents must be opposite. Therefore, the factors must be generally in the form of,

$$= \left(z_0 - 1 \cdot e^{j\omega} \right)\left(z_0 - 1 \cdot e^{-j\omega} \right)$$

Next, we must solve for ω. To do this, we must force the OI terms of the FOIL operation equal to $-z$:

$$-z_0 \cdot e^{j\omega} - z_0 \cdot e^{-j\omega} = -z_0$$

$$e^{j\omega} + e^{-j\omega} = 1$$

Here, if we divide both sides by 2, then according to Euler's identity, we have a cosine on the left-hand side of the equation. Then we can use the arccosine to solve for ω:

$$\frac{e^{j\omega} + e^{-j\omega}}{2} = \frac{1}{2}$$

$$\cos(\omega) = \frac{1}{2}$$

$$\omega = \mathrm{acos}\left(\frac{1}{2} \right) = \frac{\pi}{3}$$

Since we now have ω, we can plug back into our equation to find the roots of the numerator:

$$0 = \left(z_0 - e^{\frac{j\pi}{3}} \right)\left(z_0 - e^{-\frac{j\pi}{3}} \right)$$

$$z_0 = e^{\pm\frac{j\pi}{3}}$$

This filter has two zeros on the z-plane, but this time off of the horizontal axis, as shown in Figure 9.7.

Note that the two zero locations are conjugates of one another. This is always the case for filters if they are to have real coefficients and produce real-valued outputs, given real-valued inputs. The effect of any zero or pole must be cancelled out in the imaginary dimension by a conjugate. In the case of an odd ordered filter, then the remaining (unpaired) zero or pole must be on the horizontal axis, as we saw with the first-order LPF and HPF filters.

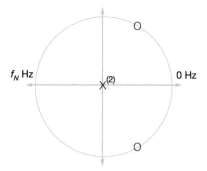

Figure 9.7

This filter has two zeros on the unit circle at $z_o = e^{\pm \frac{j\pi}{3}}$ and two poles at the origin, $z_x = 0$. The number of poles are parenthetically noted.

Alternatively, we could use the **zplane**() function to generate this plot for us, first by passing in the numerator coefficients, which of course are just the values of the IR, and the denominator coefficient, 1 (no impact). This function produces a plot very similar to Figure 9.7.

```
h=[1,-1,1];
den=1;
zplane(h, den)
```

Finally, to plot the magnitude response, we have two options. The first is to model the magnitude response analytically, and the second is to compute the frequency response with **freqz**(). Based on the frequency response and plugging in the value $\omega = \dfrac{\pi}{3}$, as determined above, the magnitude response is given by

$$|H(f)| = |H(z)|_{z=e^{\frac{j2\pi f}{fs}}} = \left| e^{j2\pi \frac{f}{fs}} - e^{j\frac{\pi}{3}} \right| \cdot \left| e^{j2\pi \frac{f}{fs}} - e^{-j\frac{\pi}{3}} \right|$$

This tells us that the magnitude is the product of the distances from the unit circle at any arbitrary frequency, f, to each of the zeros on the p/z plot. When $\dfrac{f}{f_N} = \dfrac{1}{3}$, then the magnitude is 0, and <u>none</u> of that frequency passes through the filter. The following code demonstrates both approaches.

```
% method 1 - define the magnitude response

Fs=44100;      % sampling rate
fdig = pi/3;   % notch location
f=1:Fs/2;      % frequencies to analyze
w=2*pi*f/Fs;   % omega
e=exp(1);      % e
z=e.^(j*w);    % z on the unit circle
```

9. z-Domain

```
% magnitude response
mag=abs(z-e^(j*fdig)).*abs(z-e^(-j*fdig));
plot(f, 20*log10(mag))
box on; grid on; xlabel('Freq (norm)'); ylabel('Mag (dB)');
title('Notch filter')

% method 2 - define the impulse response
h=[1,-1,1];    % freq. resp. numerator (same as IR)
den=1;         % freq. resp. denominator

% return the complex freq. resp. and corresponding freq. array
[H,F]=freqz(h, den, length(f), Fs);

hold on;
plot(F, 20*log10(abs(H)), 'r--')
```

9.6 Filter phase response

The frequency response of a digital filter is (as you now know) complex. Like any complex value, a frequency response can be thought of in terms of real and imaginary components or as magnitude and phase components. Both representations were utilized in this chapter. For example, poles and zeros were plotted on the z-plane with real and imaginary axes. And in Chapter 9.2, the magnitude of a frequency response was discussed. In this section, the other half to the magnitude response, known as the *phase response* will be considered. If the magnitude response tells us "how much" of a frequency is passed by a filter, then the phase response tells us "how long" a frequency takes to traverse a filter.

Consider the following frequency response:

$$H(f)=1+a_1e^{-j2\pi\frac{f}{fs}} \tag{9.22}$$

This filter has the following IR:

$$h[n]=[1,a_1] \tag{9.23}$$

We know that a_1 belongs in the second position (1 sample of delay) since the time-shift in (9.22) is $k = -1$. The difference equation is given by

$$y[n]=x[n]+a_1\cdot x[n-1] \tag{9.24}$$

And the transfer function is

$$H(z)=1+a_1z^{-1} \tag{9.25}$$

Therefore, the z_0 locations can be determined by solving for the root:

$$0 = 1 + a_1 z_0^{-1} \qquad (9.26a)$$

$$z_0 = -a_1 \qquad (9.26b)$$

And the z_x locations (root of the denominator) is clearly $z_x = 0$ since this is an FIR filter. To find the magnitude and phase responses, the real and imaginary parts must first be determined. To find them, first H must be converted to rectangular form:

$$H(f) = 1 + a_1 \cos(\omega) - j \cdot a_1 \sin(\omega) \qquad (9.27a)$$

$$\mathrm{Re}\{H\} = 1 + a_1 \cos(\omega) \qquad (9.27b)$$

$$\mathrm{Im}\{H\} = -a_1 \sin(\omega) \qquad (9.27c)$$

Then, to find the magnitude response:

$$|H| = \sqrt{\mathrm{Re}\{H\}^2 + \mathrm{Im}\{H\}^2} \qquad (9.28)$$

While the phase response is given by

$$\angle H = \mathrm{atan}\left(\frac{\mathrm{Im}\{H\}}{\mathrm{Re}\{H\}}\right) \qquad (9.29)$$

Interestingly, looking back at (2.1a) and (2.1b), it can be seen that the calculation for magnitude and phase are identical for a static vector as they are for a filter. For this digital filter, the phase response is

$$\angle H(f) = \mathrm{atan}\left(\frac{-a_1 \sin(\omega)}{1 + a_1 \cos(\omega)}\right) \qquad (9.30)$$

9.7 Group delay

While the phase response indicates how many cycles a particular frequency takes to traverse a filter, the interpretation can be slightly confounded by the fact that HFs naturally undergo much more rapid phase change than LFs over the same time interval. So even in a pure delay (all-pass) filter, with no distortions to magnitude or phase, the filter will exhibit a linear slope, where the steepness of the slope depends on the filter order (higher order implies a steeper slope). Phase distortions will manifest as deviations from linearity, which can be obscured when plotting the phase response since the deviations may be small compared to the steepness of the slope.

It is possible to accentuate these particular phase deviations from linearity by applying the derivative. The negative derivative of the phase response is known as the *group delay*, which indicates the rate of change of the phase at any given frequency and is defined as

$$\tau_g\left(f\right) = -\frac{d\angle H\left(f\right)}{df} \tag{9.31}$$

The group delay can be visualized this way; imagine a sine wave carrier with an envelope that smoothly increases up to a maximal point then decreases, symmetrically (perhaps a Gaussian shape). The number of samples that it takes for the peak of that digital signal to pass through a filter is the group delay for that frequency. Thinking back to a linear phase filter, the derivative of a line is a constant, which is to say that all frequencies in a linear phase filter are delayed by the same number of samples. This is how the waveshape is maximally preserved with linear phase filters. If some frequencies are delayed differently than others, then onsets and offsets, in particular, will be appreciably "smeared" out in time. Conversely, a minimum phase filter will exhibit the lowest group delay. Group delay for a linear phase and minimum phase filter (of identical order and cutoff) is shown in Figure 9.8.

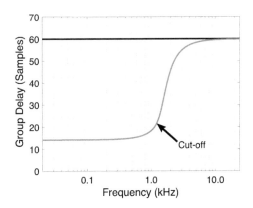

Figure 9.8

The group delay for an LPF (f_S = 48 kHz, f_{pass} = 1.2 kHz, f_{stop} = 2.4 kHz) is shown for linear phase (black) and minimum phase (gray) realizations. The linear phase realization has a constant group delay of 60/48 kHz = 1.25 ms, while the minimum phase realization has a frequency-dependent group delay ranging from 0.3 ms at 0 Hz up to 0.44 ms at 1.2 kHz.

9.8 Challenges

You are given the following IR, $h[n] = [1, -1, 0.09]$, which you implement in an FIR filter structure, whose output is given as $y[n] = h[n] * x[n]$, where $x[n]$ is the input signal.

1. What is the z-transform, $Y(z)$, of $y[n]$?
2. What is the transfer function, $H(z)$?
3. Determine the zero locations of H and plot them on the z-plane.
4. What is the frequency response, $H(f)$?
5. Calculate the gain of this filter at: $f = 0$ Hz, $f = 12,000$ Hz, and $f = 24,000$ Hz for a sample rate of $f_S = 48$ kHz.
6. Using MATLAB® or Octave, plot the magnitude response and group delay.
   a. What is the shape of this filter?
   b. What is its cutoff frequency?
   c. What is the frequency and delay (in samples) of the maximal group delay within the passband?

10

IIR filters

In Chapter 8, FIR filters (which utilize feed-forward taps) were introduced and then analyzed in the z-domain in Chapter 9. In this Chapter, IIR filters (which include feed-back taps) will be presented, along with methods for calculating their z-transform and adding poles to a p/z plot. Having the output of a filter fed back to the input first results in an IR that is infinite in length (at least in theory, if not in practice). This raises an important concern of stability, which will be examined in this chapter. But by incorporating feedback taps, an IIR filter can achieve comparable filter specifi-cations (passband ripple, rejection ratio, transition bandwidth) as an FIR filter, but with a much lower filter order, easily more than an order of magnitude. However, this comes at the cost of guaranteed phase distortion – by virtue of having an asymmetri-cal IR, an IIR filter cannot exhibit constant group delay. In this chapter, some design guides for first- and second-order IIR filters, from specification to implementation, are described along with programming examples. IIR filters are highly versatile, and the design of several filter types (shelving, HPF, LPF, BPF, etc.) are fully worked out.

10.1 General characteristics of IIR filters

By virtue of having feedback taps, differences between IIR and FIR filters show up in block diagrams, difference equations, their respective IRs, transfer functions, and p/z plots – but these differences are all inter-related. First, consider the block diagram in Figure 10.1.

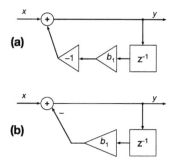

Figure 10.1

(a) A first-order IIR filter with one feed-forward tap and one feedback tap that includes a delay and a gain. (b) The negation (-1) is often denoted with a minus sign next to the accumulator. The block diagrams in (a) and (b) are equivalent.

The feedback taps will always be negated for reasons that will be discussed later. Notice in the block diagram following the accumulator, the output is split, with a copy re-entering the accumulator as a tap, but here the tap originates from the output rather than the input (the definition of a feedback tap). The feedback taps are represented in the difference equation, similar to a feed-forward tap. Starting at the accumulator, count the number of taps contributing to the output (in Figure 10.1, there are two). The first tap comes from the input, $x[n]$, and has no delay and no gain, while the second tap is fed back from the output, with a delay of one and a gain, b_1, which is then negated. The difference equation for this IIR filter is given by

$$y[n] = x[n] - b_1 y[n-1] \tag{10.1}$$

While it is clearly obvious that the immediately preceding output is directly contributing, actually all previous outputs are also indirectly contributing. To visualize this, we will next examine the impact of a feedback tap on the IR, $h[n]$. Recall that the IR is defined as the output of a filter when a delta function is the input. The IR is shown in Table 10.1, for the first five samples (column 1), the value of the corresponding input (column 2), the state of the delayed output tap (column 3), and finally the output of the filter (column 4). In the first row, there is no previous output $h[n-1]$, so this value is 0, and the output is simply the unscaled input.

Notice how the output from any given row shows up as in the subsequent row under column 3. For this particular filter, if $b_1 < 1$, as n approaches infinity, then $h[n]$ approaches 0. The general form of an IIR difference equation is a sum of all feed-forward and feedback taps, given by

$$y[n] = \sum_{k=0}^{N} a_k \cdot x[n-k] - \sum_{k=1}^{M} b_k \cdot y[n-k] \tag{10.2}$$

Table 10.1 Computing the impulse response for the IIR filter in Figure 10.1.

n	$\delta[n]$	$h[n-1]$	$h[n]$
0	1	n.d. (0)	$1 - 0 = 1$
1	0	1	$0 - b_1 \cdot 1 = -b_1$
2	0	$-b_1$	$0 - b_1 \cdot (-b_1) = b_1^2$
3	0	b_1^2	$0 - b_1 \cdot b_1^2 = -b_1^3$
4	0	$-b_1^3$	$0 - b_1 \cdot (-b_1^3) = b_1^4$
...	...	...	...
k	0	$(-1)^{k-1} \cdot b_1^{k-1}$	$(-1)^k \cdot b_1^k$

Note that all of the feedback terms are negated, as in Figure 10.1 – this is a common practice in many digital signal processors that will be discussed later. We previously used the variable N to represent the order for an FIR filter. Here, for the feed-forward portion, N will still be used to represent the maximal delay applied to the feedforward taps, while M will represent the maximal delay applied to the feedback taps. The order of an IIR filter is the maximum of N and M.

Note that in the Equation (10.2) for the summation of feedback taps, k starts at 1 as opposed to 0. A gain applied to the non-delayed output $y[n]$ acts simply as a gain, raising the overall level of the output up or down, and for this reason is usually left out of the filter coefficients and assumed to be 1.

It is important to reconsider the filter's coefficients. Whereas with FIR filters, the coefficients comprised only the set of feed-forward gains $(a_0, ..., a_N)$, with IIR filters, we also add the set of feedback gains $(b_1, ..., b_M)$. An IIR filter can be completely described by this set of coefficients.

10.1.1 Denormal numbers

As an IR tail approaches 0, its values can get very small. When they drop to this level, the digital quantization levels become a significant compared to the signal level. When this happens, the mathematical operations performed on these low-level signals are no longer linear, and these values are known as *denormal numbers*. A problem with denormal numbers is that they lead to a significant hit to performance, often running 100 times slower than normal numbers. It is typical in audio processing to handle denormal numbers with a digital noise gate/expander. These signals are already so low as to be inaudible, and so once a very low threshold is reached, we force the value to 0. This has the added computational benefit of bypassing any audio algorithms that detect silence on the input and automatically output silence, as is common in many audio effects plugins. Alternatively, denormals can be guarded against by adding a very small offset to the input (e.g., 10^{-15}), which is inaudible but well above the range of denormals.

10.2 IIR filter transfer functions

The filter described in Equation (10.1) can also be rewritten as a transfer function, in terms of the complex variable z. Recall, that from the transfer function we can directly read the filter coefficients to use in the difference equation. Just like with FIR filters, we first apply the z-transform:

$$Y(z) = X(z) - b_1 Y(z) \cdot z^{-1} \tag{10.3}$$

Next, group and then factor out the $Y(z)$ and $X(z)$ terms:

$$Y(z) \cdot \left(1 + b_1 z^{-1}\right) = X(z) \tag{10.4}$$

Finally, divide the output by the input to obtain the transfer function, $H(z)$:

$$H(z) = \frac{Y(z)}{X(z)} = \frac{1}{1 + b_1 z^{-1}} \tag{10.5}$$

Note that the denominator coefficient (b_1) in $H(z)$ has the opposite sign as the feedback coefficients in the difference equation. This is the reason that IIR filter block diagrams are implemented with the feedback taps being negated prior to the output accumulator – the benefit of this approach is that the sign of the feedback coefficients is the same in the difference equation, block diagram, and transfer function. However, in some DSP texts and software environments, this convention may not be used. A digital audio engineer must be mindful of the signs of the coefficients when working with IIR filters.

10.2.1 Programming example: first-order IIR filter

Consider the first-order IIR filter, given by the block diagram in Figure 10.2.

Problem: Determine the following:
- The difference equation,
- The pole and zero locations,
- The magnitude response,
- Group delay.

Solution: To write the difference equation, the first step is to count the number of terms contributing to the output accumulator (there are two feed-forward and one feedback taps). The no-delay tap has a gain of a_0, while the unit delay tap has a gain

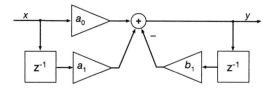

Figure 10.2

A first-order IIR filter with one delayed feed-forward tap and one delayed feedback tap.

10. IIR filters

of a_1, and the unit delay feedback tap has a gain of b_1. The difference equation is thus given by

$$y[n] = a_0 x[n] + a_1 x[n-1] - b_1 y[n-1]$$

The transfer function is determined by performing the z-transform, grouping the X and Y terms, and dividing the output by the input:

$$Y(z) = a_0 X(z) + a_1 X(z) \cdot z^{-1} - b_1 Y(z) \cdot z^{-1}$$

$$Y(z) \cdot \left(1 + b_1 z^{-1}\right) = X(z) \cdot \left(a_0 + a_1 z^{-1}\right)$$

$$H(z) = \frac{Y(z)}{X(z)} = \frac{a_0 + a_1 z^{-1}}{1 + b_1 z^{-1}} = \frac{a_0 z + a_1}{z + b_1}$$

Next, to find the z_x (pole) and z_o (zero) locations as shown in Figure 10.3, first solve for the roots of the numerator (zeros, feed-forward) and denominator (poles, feedback). To find the z_o locations, set the numerator to 0:

$$a_0 z_o + a_1 = 0$$

$$z_o = -\frac{a_1}{a_0}$$

The z_x location is non-zero since this is an IIR filter, and is found by setting the denominator equal to 0:

$$z_x + b_1 = 0$$

$$z_x = -b_1$$

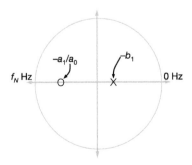

Figure 10.3

The pole and zero locations are shown for a first-order IIR filter with numerator coefficients (a_0, a_1) and denominator coefficients (1, b_1).

To find the magnitude response, start with the transfer function and evaluate on the unit circle to get the frequency response:

$$H(f) = H(z)\big|_{z=e^{j2\pi\frac{f}{fs}}} = \frac{a_0 e^{j2\pi\frac{f}{fs}} + a_1}{e^{j2\pi\frac{f}{fs}} + b_1}$$

Once we've determined the frequency response, we can find the magnitude response and group delay. For this example, select $a_0 = 0.9$, $a_1 = 0.9$, and $b_1 = 0.1$ and $f_S = 48$ kHz. Plot the frequency response derived above as well as the one calculated by software using **freqz()**.

```
% 1st order IIR filter magnitude
a0=0.9;
a1=0.9;
b1=0.1;
fs=48000;

num=[a0, a1];
den=[1, b1];

% method 1
f=0:1:fs/2;      % freq array
w=2*pi*f/fs;     % omega array
H1 = (a0*exp(j*w)+a1)./(exp(j*w)+b1);
plot(f,20*log10(abs(H1)));
grid on; xlabel('freq'); ylabel('gain (dB)')

% method 2
[H1,F]=freqz(num,den, fs, fs);
subplot(3,1,1)
plot(F,20*log10(abs(H1))); grid on
subplot(3,1,2)
plot(F, angle(H1)); grid on

% and group delay
[Grd,~]=grpdelay(num,den, fs, fs);
subplot(3,1,3);
plot(F, Grd); grid on

% p/z plot
figure;
zplane(num,den)
```

It is interesting to note that this filter exhibits *fractional delay*, in which the group delay is actually less than one sample. Also note the location of the pole and zero using **zplane()**, z_o is at $-a_1/a_0 = -0.9/0.9 = -1.0$, and z_x is at $-b_1 = -0.1$. Since this is a first-order filter, both the zero and pole are on the real axis. In Chapter 10.4, a second-order filter will be developed, which will allow for poles at frequencies other than 0 Hz or f_N Hz.

10. IIR filters

10.3 IIR filter stability

Recall the impact of the feedback gain shown in Table 10–1. It was noted that if the feedback gain, b_1, was greater than 1, then the IR would tend towards infinity as the sample index, n, increased. This is a textbook example of an instable filter. While FIR filters are guaranteed to be stable, the same is not true of IIR filters, as should be evident from this demonstration. The test for stability of an IIR filter is in examining the location of its poles. For any filter with <u>all</u> of the poles <u>within</u> the unit circle, the filter will be stable. In the case that a pole is exactly on the unit circle, the IR will be *critically stable*, and will ring at the frequency location of the pole – technically, this is no stable since the IR does not tend towards 0. Any filter with a pole outside the unit circle will exhibit instability.

10.3.1 Programming example: 3D p/z plot

If z_o is exactly on the unit circle the magnitude reaches $-\infty$ dB at the notch frequency. However, if we pull z_o off of the unit circle, then the distance between the unit circle and z_o never reaches 0; therefore, only some attenuation is applied. Conversely, a z_x results in a boost in surrounding frequencies, with the boost increasing as z_x approaches the unit circle. It could be an interesting visualization to plot the magnitude response on the entire z-plane. While the frequency response is given by the magnitude and phase on the unit circle, the poles and zeros of a filter are usually at different positions. This script allows for such visualization.

To construct a 3D visualization of the z-plane, it must be sampled. Rather than sampling the horizontal and vertical axes, we will instead sample polar form, such that the grid comprises concentric circles with equal spacing and axial lines in constant frequency spacing, using the function **meshgrid**(). Then the sampled z-plane comprises values at the intersection points of the grid. In this script, three types of filters are provided: the first is the notch filter developed in Chapter 9.5.2; the second is the first-order LPF/HPF developed in Chapter 9.5.1; the third is a second-order IIR resonator that will be discussed in Chapter 10.4. Note that for the FIR filters, the zeros pull the z-plane down, while the poles in the IIR filter push the z-plane up. In all three cases, the magnitude response is the level on the unit circle, and the phase is given by the grayscale color-coding. Each type is shown in Figure 10.4.

```
%%% z-plane grid
w = linspace(-pi, pi, 100);    % freq spacing
r = linspace(0,1,30);          % radius spacing
[wG,rG] = meshgrid(w,r);       % 2D Cartesian grid

% complex variable z (grid)
e=exp(1);
zG=rG.*e.^(j*wG);

%%% Calculate the function
%(3 different filters provided, choose one filter at a time to
plot
```

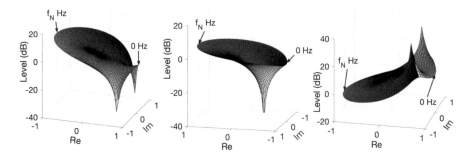

Figure 10.4

The magnitude (height) and phase (grayscale) on the z-plane is shown for three filter types: (Left) a second-order FIR notch filter is shown with z_0 directly on the unit circle; (Middle) a first-order FIR HPF is shown, with $a_1 = 0.5$; (Right) a second-order IIR resonator. In each case, following the level along the contour edge (at the radius of the unit circle) traces out the magnitude response for that filter.

```
% all 3 filters have the same sample rate)
Fs=44100;    % sample rate

% 2nd-order FIR notch filter
f= 5000;     % notch frequency
H=(zG-e^(j*2*pi*f/Fs)).*(zG-e^(-j*2*pi*f/Fs));


% 1st-order FIR LPF/HPF
a1= -0.5;
H = zG-a1;

% 2nd-order IIR resonator
f=5000;
w0=2*pi*f/Fs;
R=0.8
H = 1./(zG.^2-2*R*cos(w0)*zG+R^2);

% extract magnitude and phase
mag = 20*log10(abs(H));
ang = unwrap(angle(H));

% create rectangular grid for 'surf' function
X = rG.*cos(wG);
Y = rG.*sin(wG);
surf(X,Y,mag, ang);
xlabel('Real')
ylabel('Imaginary')
zlabel('Level (dB)')
grid on
```

10.4 Second-order resonators

By now you know that a z_o near a particular frequency produces attenuation at that frequency, while a z_x produces gain. We can imagine a pair of poles that are conjugates of one another that could be placed around the z-plane to introduce peaking gain at for a particular frequency, a filter type known also as a *resonator*. Such a filter would have a p/z plot that resembles Figure 10.5.

As the pole approaches the unit circle, the gain increases, and the range of frequencies that are affected also increases. Any frequencies that are within 3 dB of the peak frequency together comprise the bandwidth, B, of the resonator. To solve for the transfer function of this resonator with the peak at an arbitrary frequency, f_0, set the pole at that location, divide both sides of the equation by z, and subtract 1 from each side to generate one of the factors of the denominator of the transfer function.

$$z_x = A \cdot e^{j2\pi \frac{f_0}{f_s}} \tag{10.6a}$$

$$1 = A \cdot e^{\frac{j2\pi f_0}{f_s}} z^{-1} \tag{10.6b}$$

$$\underbrace{1 - A \cdot e^{\frac{j2\pi f_0}{f_s}} z^{-1}}_{\text{factor 1}} = 0 \tag{10.6c}$$

The other factor must be a conjugate, so it can be determined simply by negating the exponent:

$$\underbrace{1 - A \cdot e^{-\frac{j2\pi f_0}{f_s}} z^{-1}}_{\text{factor 2}} = 0 \tag{10.7}$$

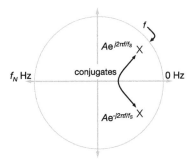

Figure 10.5

A resonator has a pair of conjugate poles that result in a peaking magnitude response at a particular frequency location, f.

These two factors are multiplied together in the denominator of H to obtain the transfer function:

$$H(z) = \frac{1}{\left(1 - A \cdot e^{\frac{j2\pi f_0}{f_s}} z^{-1}\right) \cdot \left(1 - A \cdot e^{-\frac{j2\pi f_0}{f_s}} z^{-1}\right)} \qquad (10.8a)$$

$$= \frac{1}{1 - 2A\cos\left(\dfrac{2\pi f_0}{f_s}\right) z^{-1} + A^2 z^{-2}} \qquad (10.8b)$$

From the transfer function, we can write the difference equation:

$$y[n] = x[n] - \left(-2A\cos\left(\frac{2\pi f_0}{f_s}\right)\right) \cdot y[n-1] - A^2 \cdot y[n-2] \qquad (10.9)$$

Recall that the signs of the transfer function denominator and difference equation feedback coefficients are opposite of one another. The pole radius, A, is determined by the given bandwidth, B, as

$$A = 1 - \frac{B}{2} \qquad (10.10)$$

Where the B is the normalized bandwidth (in rad), given by

$$B = \frac{2\pi f_{BW}}{f_s} \qquad (10.11)$$

And where f_{BW} is the bandwidth in Hz. However, we cannot simply assume that the pole location will correspond precisely with the peak frequency. This is because the conjugate pole is "pulling" the peak towards itself! Below $f_N/2$, this results in a shift of the actual peak frequency towards 0 Hz, and above $f_N/2$, a shift towards f_N. The impact of the conjugate pole is influenced by the pole radius. To set the pole frequency, f_x, based on our desired peak frequency, f_0, we use the following equation, given by Steiglitz [1]:

$$\cos\left(\frac{2\pi f_x}{f_s}\right) = \frac{2A}{1 + A^2}\cos\left(\frac{2\pi f_0}{f_s}\right) \qquad (10.12)$$

Finally, the amount of peaking gain depends entirely on the pole radius; the larger A is, the more gain there will be at the resonant frequency. If normalizing the gain (for example, to 0 dB) is desired, then the input (the numerator of the transfer function) must be scaled down, according to

$$a_0 = \left(1 - A^2\right) \cdot \sin\left(\frac{2\pi f_x}{f_s}\right) \tag{10.13}$$

Therefore, the filter difference equation for the resonator becomes

$$y[n] = a_0 x[n] - \left(-2A\cos\left(\frac{2\pi f_x}{f_s}\right)\right) \cdot y[n-1] - A^2 \cdot y[n-2] \tag{10.14}$$

10.4.1 Programming example: sweeping resonator

A sweeping resonator is a time-varying effect in which the resonant frequency, f_0, changes over time – in this example, f_0 increases with increasing sample index. The starting state of the sweeping resonator is centered at 100 Hz and with a bandwidth of 100 Hz. This resonator increases at a rate of 0.1 Hz per sample, or 4,410 Hz per second at a sample rate of 44.1 kHz for a duration of 2 sec. The input to the resonator is a random noise signal. Since f_0 changes every sample, the pole frequency (f_x) and normalization factor (a_0) will also be recalculated every sample.

```
T    = 2;        % duration (sec)
fs   = 44100;
f0   = 100;      % starting freq (Hz)
fBW  = 100;      % bandwidth (Hz)

B    = fBW/fs*2*pi;    % bandwidth (normalized)
A    = 1-B/2;          % pole radius

ynm1=0;
ynm2=0;

for n=1:T*fs
        % generate a sample of white noise
        x=2*(rand(1,1)-0.5);

        % sweep f0 by 0.1 Hz every sample
        f0 = f0+0.1;
        w0 = f0/fs*2*pi;

        % determine true pole location for peak resonance at psi
        wx=acos(cos(w0)*2*A/(1+A^2));

        % normalization factor (gain of x[n])
```

```
a0= (1-A^2)*sin(wx);

% implement resonator difference equation
y(n) =a0*x+2*A*cos(wx)*ynm1-A^2*ynm2;


ynm2=ynm1;   % store ynm1 into ynm2
ynm1=y(n);   % store current output into ynm1


end
sound(y, fs)
```

10.5 Biquadratic filters

Biquadratic filters, or simply "bi-quads", are second-order IIR filters comprising six coefficients (and sometimes five, but we will start with six). Bi-quads are quite versatile due to the variety of filter types that can be implemented with them – these will be covered in detail. For a review of the various filter types, refer back to Figure 1.10.

Both the transfer function and difference equation can be obtained from the direct form, as shown in Figure 10.6(a). Starting with the transfer function, we know that the feed-forward coefficients belong in the numerator, while the feedback coefficients belong in the denominator:

$$H(z) = \frac{a_0 + a_1 z^{-1} + a_2 z^{-2}}{b_0 + b_1 z^{-1} + b_2 z^{-2}} \qquad (10.15)$$

And recalling that the feedback coefficients will have their signs inverted, the difference equation can be written as

$$b_0 y[n] = a_0 x[n] + a_1 x[n-1] + a_2 x[n-2] - b_1 y[n-1] - b_2 y[n-2] \qquad (10.16)$$

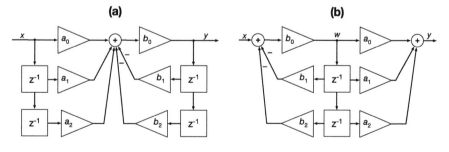

(a)   **(b)**

Figure 10.6

Block diagram for a bi-quad IIR filter in (a) direct form, and (b) canonical form.

10. IIR filters

It is somewhat unwieldy for the output, $y[n]$, to be multiplied by b_0. For this reason, it is highly common, if not universal, to scale all of the coefficients, both in the numerator and denominator, by b_0, giving

$$y[n]=\frac{1}{b_0}\cdot\left(a_0x[n]+a_1x[n-1]+a_2x[n-2]-b_1y[n-1]-b_2y[n-2]\right) \quad (10.17)$$

When we do this, then the b_0 gain can be removed from the block diagram as shown in Figure 10.6(a). By normalizing by b_0, then we are left with just five coefficients since the b_0 will always be equal to 1, but under this scenario the other coefficients must be scaled (e.g., $a_0 \rightarrow a_0/b_0$, $a_1 \rightarrow a_1/b_0$, etc.).

For canonical form, as shown in Figure 10.6(b), there will be a state variable, $w[n]$:

$$w[n]=\left(\frac{1}{b_0}\right)\cdot\left(x[n]-b_1w[n-1]-b_2w[n-2]\right) \quad (10.18)$$

The transfer function of the state variable over the input is

$$b_0W(z)+b_1W(z)\cdot z^{-1}+b_2W(z)\cdot z^{-2}=X(z) \quad (10.19a)$$

$$H_{W/X}(z)=\frac{1}{b_0+b_1z^{-1}+b_2z^{-2}} \quad (10.19b)$$

And the transfer function of the output over the state variable is

$$y[n]=a_0w[n]+a_1w[n-1]+a_2w[n-2] \quad (10.20a)$$

$$H_{Y/W}(z)=\left(a_0+a_1z^{-1}+a_2z^{-2}\right) \quad (10.20b)$$

The overall transfer function of the entire filter, $H(z)$, is given by multiplying $H_{W/X}(z)$ in Eq (10.19b) with $H_{Y/W}(z)$ in Eq. (10.20b), which is equivalent to Eq. (10.15).

10.5.1 Bi-quad design

Bi-quads can be configured as an LPF, HPF, BPF, peaking filter, notch filter, HSF, LSF, or all-pass filter using a generalized set of equations developed by Bristow-Johnson [2]. These bi-quad equations all require at least the sampling rate, f_S, the critical frequency, f_0 (for example cutoff frequency of an LPF or center frequency of a peaking filter), Q-factor, and certain filter types additionally require a gain (for example, a shelving filter).

Given f_S, f_0, and Q the digital critical frequency is first computed, followed by α, a term used in the bi-quad equations:

$$\omega_0 = \frac{2\pi f_0}{f_s} \tag{10.21a}$$

$$\alpha = \frac{\sin(\omega_0)}{2Q} \tag{10.21b}$$

Using these variables, the coefficients for several filter types can be computed.

10.5.1.1 Low-pass filter

$$a_0 = \frac{1 - \cos(\omega_0)}{2} \tag{10.22a}$$

$$a_1 = 1 - \cos(\omega_0) \tag{10.22b}$$

$$a_2 = \frac{1 - \cos(\omega_0)}{2} \tag{10.22c}$$

$$b_0 = 1 + \alpha \tag{10.22d}$$

$$b_1 = -2\cos(\omega_0) \tag{10.22e}$$

$$b_2 = 1 - \alpha \tag{10.22f}$$

10.5.1.2 High-pass filter

$$a_0 = \frac{1 + \cos(\omega_0)}{2} \tag{10.23a}$$

$$a_1 = -\left(1 - \cos(\omega_0)\right) \tag{10.23b}$$

$$a_2 = \frac{1 + \cos(\omega_0)}{2} \tag{10.23c}$$

$$b_0 = 1 + \alpha \tag{10.23d}$$

$$b_1 = -2\cos(\omega_0) \tag{10.23e}$$

$$b_2 = 1 - \alpha \qquad (10.23f)$$

10.5.1.3 Band-pass filter

For a BPF, a gain for the center frequency must be specified and converted to linear form, A.

$$a_0 = A \cdot \alpha \qquad (10.24a)$$

$$a_1 = 0 \qquad (10.24b)$$

$$a_2 = -A \cdot \alpha \qquad (10.24c)$$

$$b_0 = 1 + \alpha \qquad (10.24d)$$

$$b_1 = -2\cos(\omega_0) \qquad (10.24e)$$

$$b_2 = 1 - \alpha \qquad (10.24f)$$

10.5.1.4 Notch filter

$$a_0 = 1 \qquad (10.25a)$$

$$a_1 = -2\cos(\omega_0) \qquad (10.25b)$$

$$a_2 = 1 \qquad (10.25c)$$

$$b_0 = 1 + \alpha \qquad (10.25d)$$

$$b_1 = -2\cos(\omega_0) \qquad (10.25e)$$

$$b_2 = 1 - \alpha \qquad (10.25f)$$

10.5.1.5 Peaking filter

Unlike the BPF, in which only the pass frequencies are unattenuated, a peaking filter provides boost to a particular frequency range (depending on f0 and Q), but other frequencies are not affected. A peaking filter is shown in Figure 10.7.

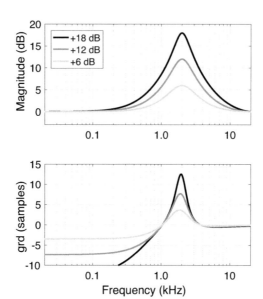

Figure 10.7

A peaking filter is shown for various gain settings. Note that increased magnitude changes (top) are accompanied with increased group delay (bottom).

$$a_0 = 1 + A \cdot \alpha \qquad (10.26a)$$

$$a_1 = -2\cos(\omega_0) \qquad (10.26b)$$

$$a_2 = 1 - A \cdot \alpha \qquad (10.26c)$$

$$b_0 = 1 + \frac{\alpha}{A} \qquad (10.26d)$$

$$b_1 = -2\cos(\omega_0) \qquad (10.26e)$$

$$b_2 = 1 - \frac{\alpha}{A} \qquad (10.26f)$$

10.5.1.6 Low-shelf filter

$$a_0 = A \cdot \left((A+1) - (A-1) \cdot \cos(\omega_0) + 2\sqrt{A} \cdot \alpha \right) \qquad (10.27a)$$

$$a_1 = 2A\left((A-1) - (A+1) \cdot \cos(\omega_0) \right) \qquad (10.27b)$$

$$a_2 = A \cdot \left((A+1) - (A-1) \cdot \cos(\omega_0) - 2\sqrt{A} \cdot \alpha \right) \qquad (10.27c)$$

$$b_0 = (A+1) + (A-1) \cdot \cos(\omega_0) + 2\sqrt{A} \cdot \alpha \qquad (10.27d)$$

$$b_1 = -2 \cdot \left((A-1) + (A+1) \cdot \cos(\omega_0) \right) \qquad (10.27e)$$

$$b_2 = (A+1) + (A-1) \cdot \cos(\omega_0) - 2\sqrt{A} \cdot \alpha \qquad (10.27f)$$

10.5.1.7 High-shelf filter

$$a_0 = A \cdot \left((A+1) - (A-1) \cdot \cos(\omega_0) + 2\sqrt{A} \cdot \alpha \right) \qquad (10.28a)$$

$$a_1 = -2A \left((A-1) - (A+1) \cdot \cos(\omega_0) \right) \qquad (10.28b)$$

$$a_2 = A \cdot \left((A+1) + (A-1) \cdot \cos(\omega_0) - 2\sqrt{A} \cdot \alpha \right) \qquad (10.28c)$$

$$b_0 = (A+1) - (A-1) \cdot \cos(\omega_0) + 2\sqrt{A} \cdot \alpha \qquad (10.28d)$$

$$b_1 = 2 \cdot \left((A-1) - (A+1) \cdot \cos(\omega_0) \right) \qquad (10.28e)$$

$$b_2 = (A+1) - (A-1) \cdot \cos(\omega_0) - 2\sqrt{A} \cdot \alpha \qquad (10.28f)$$

10.5.1.8 All-pass filter

$$a_0 = 1 - \alpha \qquad (10.29a)$$

$$a_1 = -2\cos(\omega_0) \qquad (10.29b)$$

$$a_2 = 1 + \alpha \qquad (10.29c)$$

$$b_0 = a_2 \qquad (10.29d)$$

$$b_1 = a_1 \qquad (10.29e)$$

$$b_2 = a_0 \qquad (10.29f)$$

While APFs do not affect the magnitude of a signal, they do affect the phase, which has some useful applications in audio. The location of the APF poles are dictated by the critical frequency f_0 and the amount of phase shift around f_0 is determined by the Q factor. The APF has complementing z_x and z_o, whereby every z_x within the unit circle has a corresponding z_o outside the unit circle at $z_o = 1/z_x$. As a result, the coefficients of the APF have symmetry, whereby $a_0 = b_N$, $a_1 = b_{N-1}$, $a_2 = b_{N-2}$, and so on.

The phase for the APF monotonically decreases to $-N\pi$ rad at f_N, therefore a bi-quad will have a phase of 2π at f_N. This property of APFs can be exploited by intentionally placing the phase of π rad (or 180 deg) at a particular frequency. Summing the output of the APF with the input results in a cancellation at the frequency with a phase shift of π rad, resulting in a notch filter. Similarly, a resonator can be constructed by instead differencing the output of the APF with the input. This is the basic principle behind a phaser effect, which cascades many APFs together, with the output being combined with the raw input. This results in many shifting peaks and notches that can be modulated by a low-frequency oscillator. It produces a similar effect to a comb filter, except the nulls and peaks do not have to be equally spaced.

The IR of an APF could have many different shapes, but since it has feedback, an IR tail will necessarily exist. In other words, the energy of an impulse entering the APF will be "spread out". This property has been commonly exploited by broadcast engineers to reduce transients with large crest factors, without introducing non-linearities to the signal. APFs are therefore a viable method for peak limiting, but unlike a compressor (that achieves the same end goal), an APF does not create any magnitude distortions or inject harmonics into the signal.

10.6 Proportional parametric EQ

A class of first- and second-order filters were devised by Jot, known as proportional parametric equalizers [3]. These filters are efficient (low order) and tunable equalizers whose dB magnitude response is proportional, across all frequencies, to a common dB gain control parameter, k. These filters can be cascaded together, creating a parametric EQ. The shelving filters are first-order, and the high-shelf coefficients are given by

$$a_0 = t\sqrt{k} + k \tag{10.30a}$$

$$a_1 = t\sqrt{k} - k \tag{10.30b}$$

$$b_0 = t\sqrt{k} + 1 \tag{10.30c}$$

$$b_1 = t\sqrt{k} - 1 \tag{10.30d}$$

While the low-shelf coefficients are given by

$$a_0 = tk + \sqrt{k} \tag{10.31a}$$

$$a_1 = tk - \sqrt{k} \tag{10.31b}$$

$$b_0 = t + \sqrt{k} \tag{10.31c}$$

$$b_1 = t - \sqrt{k} \tag{10.31d}$$

Where k is the gain (in linear scale), and t is a warped version of the shelf edge:

$$t = \tan\left(\frac{\pi f_0}{f_s}\right) \tag{10.32}$$

The band gain/cut filters are second-order, and the coefficients are given by

$$a_0 = t\sqrt{k} + 1 \tag{10.33a}$$

$$a_1 = -2c \tag{10.33b}$$

$$a_2 = -\left(t\sqrt{k} - 1\right) \tag{10.33c}$$

$$b_0 = \frac{t}{\sqrt{k}} + 1 \tag{10.33d}$$

$$b_1 = -2c \tag{10.33e}$$

$$b_2 = -\left(\frac{t}{\sqrt{k}} - 1\right) \tag{10.33f}$$

Where c represents the center frequency of the filter and is given by

$$c = \cos(\omega_0) = \cos\left(\frac{2\pi f_0}{f_s}\right) \tag{10.34}$$

For the second-order band filters, t now represents the bandwidth:

$$t = \tan\left(\frac{\pi f_{BW}}{f_s}\right) \tag{10.35}$$

10.6.1 Programming example: 3-band parametric EQ

A 3-band parametric EQ includes a low-shelf with selectable cutoff and gain, a center band with selectable frequency, bandwidth, and gain, as well as a high-shelf with selectable cutoff frequency and gain. These three filters will be cascaded, resulting in a fourth-order EQ. The coefficient equations for the proportional parametric EQ will be implemented in code. The EQ will be analyzed using **freqz()** with each individual filter and multiplying the results together. In this example, we will apply a bass boost of +4 dB with an edge of 100 Hz, a narrow cut of 3 dB at 500 ± 100 Hz, and a slight boost of +1 dB above 4 kHz. The magnitude response of this three-band EQ is shown in Figure 10.8.

```
%% proportional parametric equalizer
fs=44100;

% low-shelf
w_lo=2*pi*100/fs;      % 100 Hz
k_lo=db2mag(4);        % +4 dB
t_lo = tan(w_lo/2);

a0_lo=k_lo*t_lo+sqrt(k_lo);
a1_lo=k_lo*t_lo-sqrt(k_lo);
b0_lo=t_lo+sqrt(k_lo);
b1_lo=t_lo-sqrt(k_lo);

% mid band
c=cos(500*2*pi/fs);    % 500 Hz center
BW=2*pi*100/fs;        % 100 Hz bandwidth
k_mid=db2mag(-3);      % -3 dB
t_mid = tan(BW/2);

a0_mid=t_mid*sqrt(k_mid)+1;
a1_mid=-2*c;
a2_mid=-(t_mid*sqrt(k_mid)-1);
b0_mid=t_mid/sqrt(k_mid)+1;
```

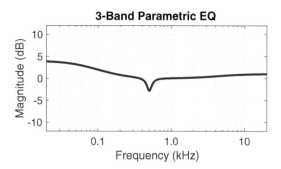

Figure 10.8

Using the proportional parametric EQ coefficients for low-shelf, band pass/stop, and high-shelf, a three-band parametric EQ was created.

```
b1_mid=-2*c;
b2_mid=-(t_mid/sqrt(k_mid)-1);

% hi-shelf
w_hi=2*pi*4000/fs;    % 4 kHz
k_hi=db2mag(1);       % +1 dB
t_hi = tan(w_hi/2);

a0_hi=sqrt(k_hi)*t_hi+k_hi;
a1_hi=sqrt(k_hi)*t_hi-k_hi;
b0_hi=sqrt(k_hi)*t_hi+1;
b1_hi=sqrt(k_hi)*t_hi-1;

% analyze coefficients
[H_HS, F] = freqz([a0_hi a1_hi], [b0_hi b1_hi], fs, fs);
H_LS = freqz([a0_lo a1_lo], [b0_lo b1_lo], fs, fs);
H_BP = freqz([a0_mid a1_mid a2_mid], [b0_mid b1_mid b2_mid], fs,
fs);

% plot magnitude response
semilogx(F, 20*log10(abs(H_LS.*H_BP.*H_HS)))
grid on;
axis([20 20000 -12 12]);
```

10.7 Forward-reverse filtering

Since the IR of a recursive filter is asymmetric by definition, then phase distortion is an inevitability. However, it is possible to remove the phase distortion of an IIR filter, as long as the conditions of causality are removed. A signal passing through a filter results in the signal being convolved with the filter's IR. If the output of the filter were then reversed and passed <u>again</u> through the filter, then the IR is again convolved with the signal, but this time in the opposite direction, resulting in a symmetric IR, hence linear phase. Time-reversing the output once again restores it to its original directionality. This process is known as forward-reverse filtering and results in constant group delay of 0 across all frequencies. The overall process can be visualized according to Figure 10.9, with an intermediate state that is the output of the first filter, $w[n]$.

Figure 10.9

The process of forward-reverse filtering results in zero phase distortion and a squaring of the magnitude response. The process involves a signal being passed through a filter, then time-reversed and passed through the same filter again, and finally time-reversed once again.

The additional effect of running a signal through a filter twice is that the magnitude response becomes squared (or in dB terms, multiplied by two) – this results in a transition band with a steeper slope, and a stopband with greater attenuation; although, any passband ripple will also be enhanced, as well. Equally stated, we could instead time-reverse the filter's IR, resulting in the following:

$$y[n] = h[-n] * (h[n] * x[n])$$ (10.36)

The z-transform of $h[-n]$ is $H(z^{-1})$. Applying the z-transform to Eq. (10.36), we get:

$$Y(z) = H(z^{-1}) \cdot H(z) \cdot X(z)$$ (10.37)

To obtain the magnitude response, evaluate on the unit circle, and applying the theorem from equation (2.8) regarding the multiplication of a complex value with its conjugate:

$$Y(\omega) = H(-\omega) \cdot H(\omega) \cdot X(\omega) = |H(\omega)|^2 \cdot X(\omega)$$ (10.38)

10.8 Challenges

1. Consider a resonator filter with a z_x at 12 kHz and a radius of $A = 0.5$. This filter is operating at a sample rate of $f_S = 48$ kHz.
   a. Write the equation for this filter using Equation (10.8).
   b. What is the magnitude of this filter at 12 kHz?
   c. What is the magnitude at 0 Hz?
2. Use the Jot proportional parametric EQ coefficient equations to design a low-shelf filter with a sample rate of 44.1 kHz and a cutoff frequency of 1.1025 kHz.
   a. What are t and k?
   b. Calculate the feed-forward (a) and feedback (b) coefficients.
   c. Sketch the p/z plot labeling the poles with 'x' and the zeros with 'o'
   d. Write the transfer function for this filter.
3. Use the Bristow-Johnson bi-quad coefficient equations to design a DC blocking filter with a sample rate of 48 kHz and a cutoff of 48 Hz.
   a. What are ω_0 and α (assume a Q-factor of 0.5)?
   b. Calculate the feed-forward (a) and feedback (b) coefficients.
   c. Sketch the p/z plot.
   d. Write the difference equation of this filter.

10.9 Project – resonator

The **filter()** and **filtfilt()** commands can be used to process an audio signal (or any signal) through a filter with specified numerator and denominator coefficients (in decreasing powers of z). For example, given the following transfer function:

$$H(z) = \frac{a_0 + a_1 z^{-1}}{1 + b_1 z^{-1}}$$

The numerator and denominator coefficients are:

```
num = [a₀ a₁];
den = [1 b₁];
```

If we want to filter one second of white noise, x, that would be accomplished by:

```
x=2*rand(1,fs)-1;
y=filter(num, den, x);
% or with zero phase distortion
y=filtfilt(num, den, x);
```

Given the numerator and denominator, the frequency response can be computed using the **freqz()** command, which also takes four arguments: numerator, denominator, the number of points to evaluate around the top half of the unit circle, and sample rate (optional). Note that the negative half will be symmetric to the positive half. The output variables of **freqz()** are the complex frequency response, and a frequency vector of that corresponds to the length of the frequency response.

Some additionally useful MATLAB®/Octave commands for working with digital filters are listed below:

- **roots()**: computes the roots of a polynomial from a list of its coefficients – useful to determine the pole and zero locations from the transfer function
- **poly()**: computes the coefficients of a polynomial from its roots – useful to determine the transfer function numerator and denominator, given the zero and pole locations
- **zplane()**: plots the pole and zero locations on the z-plane, given either the numerator and denominator coefficients or the pole and zero locations

In this project, you will code a second-order resonating filter using Equations (10.10) through (10.14). The procedure involves first specifying a bandwidth and center frequency to determine the pole radius. Next, you will determine your actual pole angle (recall, this will differ slightly from the center frequency angle) and normalized amplitude. With these variables calculated, you can generate your difference equation. Then determine the transfer function of this filter, and use the numerator and denominator coefficients to process audio using the **filter()** and **filtfilt()** functions.

Assignment

1. Devise the difference equation and transfer function for this resonator.
2. Use the **roots()** function to find the pole and zero locations.
3. Use the **zplane()** function to plot the poles and zeros on the z-plane.
4. Using **freqz()**, plot the magnitude (in dB) and phase response (in radians) with a logarithmic frequency axis.

5. Devise a <u>new</u> resonator to boost frequencies at 2 kHz with a bandwidth of 1 kHz.
   a. Plot the magnitude response, phase response, and p/z locations.
   b. Using **filter**() <u>and</u> **filtfilt**() Apply this filter to a sound file – do you hear a difference between these two?

Bibliography

[1] Steiglitz, K. "A Note on Constant-Gain Digital Resonators." *Computer Music Journal*, Vol. 18, No. 4 (1994): 8–10.
[2] Bristow-Johnson, R. "Cookbook Formulae for Audio EQ Biquad Filter Co-efficients." *Online Resource*, available: http://music.columbia.edu/pipermail/music-dsp/2001-March/041752.html
[3] Jot, J.M. "Proportional Parametric Equalizers – Application to Digital Reverberation and Environmental Audio Processing." Presented at the 139th Convention of the Audio Engineering Society, New York, 29 Oct – 1 Nov, 2015. Paper 9358 (8 pgs.)

Impulse response measurements

You may be familiar already with the concept of capturing the IR of an audio *device under test* (DUT). The DUT itself could be acoustic with an acoustic stimulus (such as the room response to a pistol blank), an acoustic DUT with an electrical stimulus (such as a loudspeaker), an electrical DUT with an electrical stimulus (guitar pedal or outboard gear), or even a digital system and stimulus, like an effects plug-in. In each case, there are some considerations to take into account. For example, with an acoustic impulse, there is a trade-off between providing power to the impulse by increasing the impulse's width versus extending the high-frequency range by decreasing the impulse's width.

The level of an impulse can be difficult to relate to steady-state signals. While the peak of an impulse or steady-state signal is easily defined, this is not a good intensity measure of comparison since steady-state signals are symmetrical in the positive and negative phase, while impulse sources typically are not. From a neurophysiological perspective, a click and a steady-state signal could be considered to have comparable levels when the auditory nerves respond similarly. This is known as *peak equivalent* level and requires an impulse to have an amplitude three times greater than a steady-state signal to achieve the same perceptual level [1].

Another problem with IR recordings can be the level of ambient noise, which can take the form of electrical or mechanical hum, or unwanted (but perhaps unavoidable) acoustic sounds. If the noise is sufficiently random (with a Gaussian-like pdf – see Chapter 1.1 for a refresher on these topics), then the level of noise can be reduced through the process of averaging. If we record multiple IRs, then the reproducible part is expected to be more or less the same between these recordings – namely, the response of the DUT to the impulse. On the other hand, the random parts of the recording (that is, the noise) are not expected to be the same from one recording to the next; therefore, upon averaging, the averaged level of the noise is expected to decrease, compared to the level of any individual recording.

11.1 Noise reduction through averaging

One hurdle when capturing IRs is the presence of background noise levels. On one hand, a greater stimulus level means better SNR. However, stimulus levels may be limited due to the range of inputs that are within the linear region of the audio DUT. If the stimulus level cannot be increased, and the background noise cannot be decreased, then it becomes impossible to improve the SNR of a single-shot IR recording, which can become problematic.

If it can be assumed that the noise is truly random, then the level of acquired noise will be reduced through the process of averaging, as shown in Figure 11.1. That is because when two uncorrelated noise signals are added together, the RMS level of their sum increases by only 3 dB. Then, when obtaining the average (dividing by two), the RMS level decreases to –6 dB, putting the average at a –3 dB compared to the level either of the two signals being averaged together.

If the random noise, η, also exhibits a near-zero DC term, then the RMS of the averaged noise, $\bar{\eta}$, comprising N individual IRs can be written as

$$RMS(\bar{\eta}) = \frac{1}{\sqrt{N}} RMS(\eta) \qquad (11.1)$$

Interpreted in terms of dB, the RMS of averaged noise decreases by –3 dB for each doubling of N. However, the fact that IRs cannot overlap puts a limit on the number

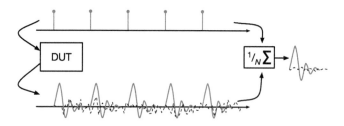

Figure 11.1

To reduce the noise (dashed) of an IR recording, many IRs (solid) can be collected then averaged together. The stimulus spacing (top) must be far enough apart to prevent overlapping of IRs during acquisition (bottom). Notice that following averaging, the level of the noise decreases.

11. Impulse response measurements

of IRs that can be acquired for a given acquisition period. For example, to achieve 30 dB of noise reduction through averaging would require over 1,000 stimuli. Since the tail of an IR contains much of the character of the IR, it should be preserved as much as feasible. Luckily, a rapid IR acquisition technique was devised using a highly jittered stimulus sequence known as a *maximum length sequence* (MLS), which allows for the capture of overlapping IRs that can then be unentwined into a single, averaged IR.

11.2 Capturing IRs with MLS

The sequence of stimulus locations in Figure 11.1 is said to be *isochronic*, the amount of space between the stimuli is equal. A different ordering of stimuli results in a highly jittered sequence, known as an MLS. An MLS is simply a temporal sequence of ones (which represent stimulus locations) and zeros (which represent silence).

If an MLS is passed through an audio DUT, an IR is produced at each of the stimulus locations. However, since the length of the IR is often longer than the interval between successive stimuli, overlapping of the IRs will occur. Despite this, when the output of the DUT driven by an MLS is circularly correlated with that same stimulus sequence, then the overlapping IRs become untwined and averaged, resulting in a single IR. This is because of a particular feature of an MLS, which is that when it is *circularly* correlated with itself ($s_{ML}[n]$), the resultant sequence is a dirac delta:

$$s_{ML}[n] \circledast s_{ML}[n] = \delta[n] \tag{11.2}$$

If a DUT is stimulated with $s_{ML}[n]$, then while the individual responses to each stimulus are $h[n]$, the IRs each overlap, giving an output $y[n]$.

$$y[n] = s_{ML}[n] * h[n] \tag{11.3}$$

If the output is then circularly correlated with the stimulus, then we get:

$$y[n] \circledast s_{ML}[n] = (s_{ML}[n] * h[n]) \circledast s_{ML}[n] = h[n] \tag{11.4}$$

This process of capturing the IR of an audio DUT at very high rates by correlating the stimulus with the output of the DUT is shown in Figure 11.2.

Interestingly, the spectrum of an MLS is white, except for the DC component, which is very nearly 0 (especially at long sequence lengths). Since the stimulus is not a transient and exhibits a 0 dB crest factor, it is able to deliver a higher amount of energy to the DUT, which increases the level of the stimulus, and therefore the SNR. Furthermore, since several separate IRs combine into a single IR, noise reduction is achieved through averaging, thus improving SNR by decreasing the noise level.

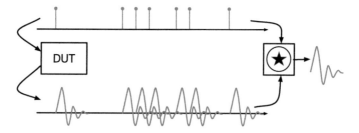

Figure 11.2

Passing an MLS through a DUT, results in an output in which IRs are often overlapping. Circular correlation of the output with the input stimulus results in a solitary IR. This is a rapid way to acquire several IRs and take advantage of noise reduction through averaging.

11.2.1 MLS limitations

The delta that is produced when auto-correlating an MLS has a length of 2^N-1. For this reason, this is the longest length of an IR that can be captured by this same method. If the IR is longer than this length, then the tail of the IR will actually wrap around and appear at the beginning of the IR, a phenomenon known as *time aliasing*. Therefore, the order of the MLS must be chosen to ensure that IR can be fully captured to prevent time aliasing.

One other limitation with the MLS method is when a DUT contains nonlinearities, which results in a distortion artifact known as "distortion peaks". These appear in the IR in regularly spaced intervals, and introduces a crackling noise, which cannot be reduced through averaging. In the case of weakly non-linear DUTs, another method is available that better handles these distortions, and beyond simply removing them, can actually characterize the distortion. This method, involving an exponential swept-sine stimulus, is covered in Chapter 11.3.

11.2.2 Example: maximum length sequence

Consider a second order ($N = 3$), with length $2^3 - 1 = 7$. The difference equation that describes the generation of an MLS of this length is

$$s_{ML}[n] = s_{ML}[n-1] \oplus s_{ML}[n-3] \tag{11.5}$$

Here, the exclusive-or (XOR) operator, represented by the encircled plus sign, acts as a modulo-two operator, whereby $0 \oplus 0 = 0$, $0 \oplus 1 = 1$, $1 \oplus 0 = 0$, and $1 \oplus 1 = 0$. The block diagram for this difference equation is shown in Figure 11.3. Only a unique, looping sequence of numbers is possible for a given order. Changing the initial states of the delays simply results in a different starting position of that sequence. So long as the delay states are not all zero-valued, the sequence is guaranteed to continue. The filter is guaranteed to have at least one non-zero state as long as the sequence length does not exceed 2^N-1, which represents the maximal length of the sequence.

11. Impulse response measurements

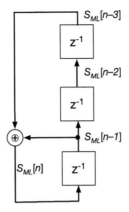

Figure 11.3

An order $N = 3$ MLS has a length of 7 values. The delay elements should not be cleared but rather given an initial state comprising 1s and 0s. The XOR summation then also produces a 1 or 0 that is fed back to the input.

11.2.3 Circular convolution/correlation

Circular convolution (or correlation) can be thought of as linear convolution (or correlation) where one of the sequences repeats itself forever in both directions. In other words, one of the sequences is periodic with its length, N. A sequence, x, can mathematically be made periodic by taking the modulus of the sample index with its length, N, thus yielding

$$x'[k] = \sum_{m=-\infty}^{\infty} x[k-mN] \tag{11.6}$$

Then if a signal, $h[n]$, is convolved or correlated with $x'[n]$, h is said to have been "circularly" convolved (Equation 11.7a) or correlated (11.7b) with x, given by

$$x[n] \circledast h[n] = \sum_{k=0}^{N} h[k] \cdot \left(\overbrace{\sum_{m=-\infty}^{\infty} x[n-k-mN]}^{x'[n-k]} \right) \tag{11.7a}$$

$$x[n] \circledast h[n] = \sum_{k=0}^{N} h[k] \cdot \left(\underbrace{\sum_{m=-\infty}^{\infty} x[n+k-mN]}_{x'[n+k]} \right) \tag{11.7b}$$

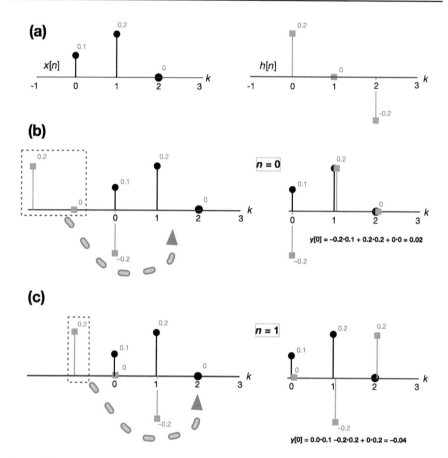

Figure 11.4

(a) Two different signals, x and h, to be <u>circularly</u> correlated together. In this example, x will be considered as the periodic one. (b) The parts of h landing on k = -1 and k = -2 can be thought of as being wrapped around to the front, as in a circular buffer. (c) As the shift increases to n = 1, there is more overlap of x and h, so only one value is wrapped. The third and final shift would show complete overlapping with no wrapping.

Since x' is periodic *ad infinitum*, the result of the circular convolution is also periodic with a length equal to the smaller of the lengths of x and h. In practice, the shorter sequence is zero-padded to the length of the longer sequence to prevent truncation. Visually, the effect of circular convolution might be even easier to understand. Figure 11.4 shows a few steps of a circular correlation example.

11.3 Capturing IRs with ESS

A *sine-sweep*, sometimes called a chirp, is a newer but common method for capturing the IR of an audio DUT. A specific variant of a sine-sweep, known as the *exponential sine-sweep* (ESS), allows for capture on a non-linear DUT, and is able

11. Impulse response measurements

to also extract IRs for the higher harmonics of a non-linear DUT (in addition to the linear IR), a technique that will be explored in Chapter 14. The idea behind any sine-sweep is to continuously drive the DUT with a sinusoid that changes frequency over time. A sine-sweep is defined by its starting and ending frequencies (f_a and f_b), its duration (T), sample rate, and the rate of increase of frequency, or *sweep rate* (L). Therefore, a sine-sweep stimulus can be designed to have large bandwidth, and sinusoids innately have low crest factors, making them an efficient stimulus for bringing almost any DUT into a steady state.

The rate of increase, L, of a sine-sweep can vary; typical rates are linear, quadratic (convex or concave), and exponential (sometimes referred to as logarithmic or geometric). For example, if the frequency is increasing linearly, then the duration of time between f_a and $f_a + 1$ Hz is the same as the duration from $f_b - 1$ Hz to f_b. In contrast to a linearly evolving sine-sweep is the ESS. The ESS perceptually sounds linear to the ear since we hear frequency in ratios of 2, and the ESS also increases in frequency at the same rate. The frequency changes of an octave occur over constant time intervals, such that the duration of time from f_a to $2f_a$ is equal to the duration from $f_b/2$ to f_b (and this ratio also holds for every octave span within the ESS stimulus).

The IR is obtained by driving a DUT with an ESS stimulus and post-processing the response. Somewhat similar to the MLS technique, in which a delta is obtained through circular auto-correlation of the stimulus sequence, a delayed delta can be obtained from an ESS by convolving the stimulus (s_{ESS}) with an <u>inverse</u> of itself (s_{ESS}^{-1}).

$$s_{ESS}[n] * s_{ESS}^{-1}[n] = \delta[n - T \cdot f_s] \tag{11.8}$$

If a DUT with IR $h[n]$ is stimulated with $s_{ESS}[n]$, then an output $y[n]$ is given by

$$y[n] = s_{ESS}[n] * h[n] \tag{11.9}$$

If the output is then convolved with the inverse ESS signal, then the following is obtained:

$$y[n] * s_{ESS}^{-1}[n] = (s_{ESS}[n] * h[n]) * s_{ESS}^{-1}[n] = h[n - T \cdot f_s] \tag{11.10}$$

This process of capturing an IR using an ESS is shown in Figure 11.5.

One slight caveat, taking the inverse of a signal can result in division by numbers close to (or at) zero, causing instability, especially in the frequency regions $<f_a$ or $>f_b$. In practice, s_{ESS}^{-1} is band-limited (simply by band-pass filtering) to frequencies between f_a and f_b, and as a result, the delta that is produced in Equation (11.8) will be band-limited as well, resembling a symmetric FIR pulse with pre-ringing. For these reasons, it is good practice to use as large bandwidth as is feasible for your application to reduce these limitations.

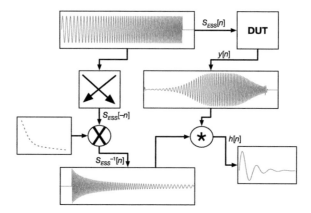

Figure 11.5

IR capture using the ESS stimulus.

11.3.1 ESS stimulus generation

An ESS is flat-top in the time domain, as shown in Figure 11.6(a), so the amplitude level at each frequency is constant. But recall that the ESS has an exponentially increasing sweep rate, L. And since the ESS "spends more time" in LFs compared to HFs, it actually has a sloped magnitude spectrum that rolls off at a rate of –10 dB/dec, as shown in Figure 11.6(b). Additionally, the ESS has a low group delay for LFs and a large group delay for HFs. Since LFs are present early in the signal and HFs are present late in the signal, the group delay of an ESS increases exponentially, as shown in Figure 11.6(c). Note that the horizontal axes in Figures 11.6(a) and (b) are logarithmic.

The magnitude of the ESS decays at a rate of –10 dB/dec, or –3 dB/oct. For a starting frequency of f_a, the magnitude, A, should be $1/\sqrt{2}$ at $2 f_a$ and $1/2$ at $4 f_a$. Therefore, in general, the magnitude is given by

$$A(f) = \sqrt{\frac{f_a}{f}} \qquad (11.11)$$

The group delay, τ_g, increases logarithmically with frequency, and should have an initial value (at f_a) of 0 s (indicating that the argument of the log should be 1). At f_b, the group delay should be exactly T s (ergo, the argument of the log should be equal to the base of the log). If the argument of the log at $f = f_b$ is f_b/f_a, then the base of the log should be the same. A change of base for a log can be achieved by dividing one log operation by the log of the desired base. Therefore, the group delay expression is given by

$$\tau_g(f) = T \cdot \frac{\log\left(\dfrac{f}{f_a}\right)}{\log\left(\dfrac{f_b}{f_a}\right)} = L \cdot \log\left(\frac{f}{f_a}\right) \qquad (11.12)$$

11. Impulse response measurements

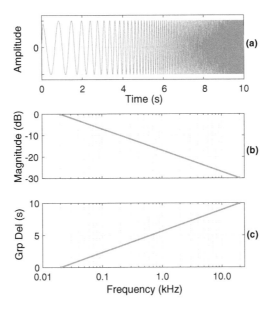

Figure 11.6

(a) An ESS starting at 20 Hz and increasing to 20.48 kHz (10 octaves) over 10 seconds, or exactly 1 octave per second – note: ESS is scaled for visualization. (b) The magnitude of the ESS rolls-off by –10 dB/dec. (c) The group delay is 0 s at 20 Hz, and increases exponentially (or linearly on a logarithmic plot) to 10 s at 20.48 kHz.

Where the sweep rate, L, is given by

$$L = \frac{T}{\log\left(\dfrac{f_b}{f_a}\right)} \tag{11.13}$$

Finally, the ESS stimulus, s_{ESS}, is given in terms of the sample index, sample rate, sweep rate, and start frequency as

$$s_{ESS}[n] = \sin\left(2\pi f_a L \cdot \left(e^{\frac{t}{L}} - 1\right)\right) \tag{11.14}$$

Careful selection of the ESS duration, T, is important to ensure that the stimulus ends at a zero-valued sample, crossing from negative to positive. This prevents a discontinuity at the end of the stimulus, allows the ESS to end on a complete cycle, and also helps improve the estimation of DUT phase. Fortunately, by rounding the sweep rate, a slightly modified version of the ESS, known as the "synchronized" ESS can be generated [2]. This is done by providing the desired duration $\hat{T}$, which approximates the duration of s_{ESS}. A synchronized sweep rate, $\hat{L}$, is given by

$$\hat{L} = \frac{1}{f_a} \text{round}\left(\hat{T} \cdot \frac{f_a}{\log\left(\dfrac{f_b}{f_a}\right)} \right) \tag{11.15}$$

Then the synchronized sweep rate can be substituted into Equation (11.14) in the place of L to generate the synchronized sine sweep.

11.3.2 Inverse ESS generation

The inverse ESS, s_{ESS}^{-1}, can be generated by time reversing the stimulus, s_{ESS}, and applying a magnitude correction, in the form of an amplitude envelope. The phase of the ESS stimulus is "undone" by time reversing it; this has the effect of putting f_b at 0 s group delay, and f_a at T s group delay. Convolution of s_{ESS} with s_{ESS}^{-1} results in addition of their respective group delays, which is a constant equal to T (recall, a constant group delay, in which all frequencies appear simultaneously, corresponds to a delayed impulse). Next, in order to correct for the sloping magnitude of s_{ESS}, an amplitude envelope with an inverted pink slope is applied, in which the level increases at a rate +3 dB/oct, or +10 dB/dec. The inverse of the ESS shown in Figure 11.6 is shown in Figure 11.7.

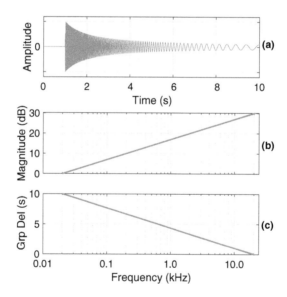

Figure 11.7

(a) An inverse ESS, s_{ESS}^{-1}, starting at f_b = 20.48 kHz and decreasing to f_a = 20 Hz over T = 10 s (note: scaled for visualization). (b) The magnitude increases by +3 dB/oct to counter the magnitude of the stimulus. (c) The group delay is 10 s at 20 Hz, and decreases exponentially (or linearly on a logarithmic plot) to 0 s at 20.48 kHz to counter the group delay of the stimulus.

11. Impulse response measurements

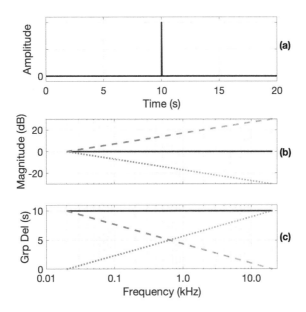

Figure 11.8

(a) The output of convolving s_{ESS}^{-1} with s_{ESS} is a delayed delta. The magnitude (b) and group delay (c) are shown in dotted gray for s_{ESS}, dashed gray for s_{ESS}^{-1} and in solid black for the result of their convolution.

The magnitude and phase responses of the inverse signal are complementary to those of the stimulus. Following convolution of s_{ESS}^{-1} with s_{ESS}, the group delays of the two signals become summed, while the magnitudes become multiplied (note, if the magnitudes are in dB then they can be summed due to properties of logarithms). These operations result in a constant group delay and a flat-spectrum magnitude response, as shown in Figure 11.8. The length of the output of the convolution is approximately $2T$ (minus a single sample), with the delta appearing at T.

11.3.3 Noise reduction

When recording the IR of a DUT, it is preferable to reduce the noise that may be recorded by the microphone. Ambient noise can be reduced by taking the recordings in a quiet environment. Further hardware considerations, such as a clean power supply or using balanced analog lines can also reduce unwanted noise. With synchronous averaging of several ESS stimuli, noise can be reduced through the process of averaging, as described in Chapter 11.1. Additionally, with the ESS IR capture technique and use of oversampling, the noise floor can be reduced even further.

If we consider the addition of noise and adapt Equation (11.10), the IR of a system with additive noise, η, is:

$$\left(s_{ESS}[n] * h[n] + \eta[n]\right) * s_{ESS}^{-1}[n] = h\left[n - T \cdot f_s\right] + \eta[n] * s_{ESS}^{-1}[n] \qquad (11.16)$$

It can be seen that the noise component is filtered by the inverse ESS, which is to say that the magnitude of the inverse ESS shapes the magnitude of the noise. If you refer back to Figure 11.7(b), you will see that the magnitude spectrum of the inverse ESS has a low-cut, high-boost characteristic, which will be imparted on to the noise. This, in turn, shapes the magnitude response of the noise to have attenuated LF levels and boosted HF levels, relative to the sampling rate. The "cross-over" point from cut to boost is at $f_S/4$. If a high enough sampling rate is used, then a significant amount of noise can be removed through low-pass filtering for $f > f_b$, such that the SNR improves by 3 dB for every doubling of f_S or every halving of f_b.

11.3.4 Programming example: noise reduction with ESS

The effects of noise reduction by increasing the sampling rate is demonstrated in this example. Let's consider an ESS from 20 Hz to 20 kHz with a duration of 2 s (followed by a silent period of 0.1 s to allow for ring out of the DUT). In this simulation, the DUT will simply be an LPF model with a cutoff of 5 kHz. We will create multiple ESS at various commonly used audio sampling rates.

To estimate the noise signal, we will drive the DUT twice, and each time a new random noise signal will be generated and added to the response. The average of these two responses is an estimate of the signal, while the difference between these two responses is an estimate of the noise. The RMS of the estimated noise will be retained for each sampling rate and compared as a dB ratio against the baseline (lowest) sampling rate, 44.1 kHz. The ESS generation and IR extraction are performed with functions built-in to MATLAB®, namely **sweeptone**() and **impzest**(), respectively.

```
fa = 20;
fb = 20000;
T = 2;
sil = 0.1;
n_rms = [];

for fs = [44100, 48000, 96000, 192000]

    % ESS
    ESS = sweeptone(T,sil,fs,'SweepFrequencyRange',[fa fb]);

    % simulated DUT
    [DUT_a, DUT_b] = butter(2,5000/(fs/2));

    % Drive DUT twice, add random noise
    y_1 = filter(DUT_a, DUT_b, ESS+0.1*randn(size(ESS)));
    y_2 = filter(DUT_a, DUT_b, ESS+0.1*randn(size(ESS)));

    % Extract IR
    ir_1=impzest(ESS, y_1);
    ir_2=impzest(ESS, y_2);

    % normalize
    ir_1 = ir_1/max(abs(ir_1));
    ir_2 = ir_2/max(abs(ir_2));
```

```
    % LPF IRs at fb
    [LPF_a, LPF_b]=butter(2,fb/(fs/2));

    ir_1f = filter(LPF_a, LPF_b, ir_1);
    ir_2f = filter(LPF_a, LPF_b, ir_2);

    % plot IRs
    hold on;
    plot((0:50)/fs, (ir_1f(1:51)+ir_2f(1:51))/2);

    % estimate noise levels
    n = ir_1f-ir_2f;
    n_rms = [n_rms rms(n)];

    % db difference of noise levels compared to 44.1 kHz
    if (length(n_rms)>1)
        20*log10(n_rms(1)/n_rms(end))
    end
end
end
```

Note the small improvement from 44.1 kHz to 48 kHz (around 0.5 dB). At 96 kHz, or more than a doubling from 44.1 kHz, we see an improvement of slightly more than 3 dB, and at 192 kHz, or just more than two doublings from 44.1 kHz, an improvement of more than 6 dB can be observed.

11.4 Challenges

1. You want to take an IR measurement with a pistol blank that has a peak-equivalent SPL level of 90 dB in an environment that has an ambient noise floor of 50 dB-SPL (RMS). How many IR captures are required to obtain a SNR of 60 dB through the process of averaging?
2. Generate an order $N = 3$ maximum length sequence, which will have a length of 7 values.
3. Perform circular auto-correlation of the MLS generated in Challenge #2.
4. Consider an ESS with $f_a = 62.5$ Hz, $f_b = 16,000$ Hz, $f_S = 48$ kHz, and a desired duration of $T = 8$ s.
   a. When does the frequency $f = 1$ kHz occur?
   b. What is the synchronized sweep rate?
   c. What is the actual duration of the ESS?
5. You are using an ESS to drive a loudspeaker in order to capture a room IR up to a frequency of 20 kHz. What is the best sampling rate to use, 44.1 kHz, 48 kHz, or 96 kHz, and why?

11.5 Project – room response measurements

MATLAB® has an application called the "Impulse Response Measurer" (part of the Audio Toolbox) that can be used to capture the IR of an acoustic or audio system, such as an audio I/O device or a room. Two capture techniques are available, using either the ESS or MLS stimulus. To capture the IR of your room at your listening location, simply grab a microphone, and place it on a stand near your sitting area. Note that both the frequency response of the microphone as well as the loudspeakers will color the acquired IR, so it would be helpful to utilize a reference microphone and loudspeaker, if they are available.

Room acoustics are heavily influenced by both the nodes and anti-nodes of the room modes, as well as interference from reflections off of large surfaces. Both of these phenomena are position-specific and can change subtly or dramatically over a difference in listening positions of less than a foot. For this reason, it is recommended to capture IRs at multiple locations within the realm of your listening position. Try to capture positions above/below, in front/behind, and left/right in addition to the typical position. These IRs can then all be averaged together to create a composite IR that is not specific to any single location.

Assignment

1. To open the application, simply enter impulseResponseMeasurer into the command window. You will, by default, get to the options for MLS acquisition. To the very left of the toolbar, you will see information about your sound card – update your device input (recorder) and output (player) channels are correct. It is easy to check for full loop-back using the "Monitor" feature towards the center-right of the toolbar, which allows you to preview different channels on your sound card.

2. Next, update the parameters for MLS capture, using the following:
   - Number of Runs: 5
   - Duration per Run: 5 (s)
   - Excitation Level: –6 (dB-FS)

3. Perform a capture – the IR will be held in memory, and visible in the Data Browser to the left.

4. Next, switch the capture method to ESS. Verify that the number of runs, duration, and excitation level matches the parameters above. Perform another capture.

5. Are there any differences between the IRs capture with MLS and ESS stimuli?
   a. If so, describe them.
   b. What do you think is the reason for these differences, or lack thereof?

6. Repeat steps 2 and 3 at six additional locations around your listening position using the capture method of your choice.

7. Repeat step 6 for your other loudspeaker (if you were using Left before, switch to Right)

8. Select your 14 IRs and Export (far right of the toolbar) them to the workspace. Average these together and plot the combined IR. Access each individual IR by using a command such as:

```
irdata_xxxxxx.ImpulseResponse(1).Amplitude;
```

    a. Where 'xxxxxx' is a timecode, and (1) represents the index of each capture that you performed.

Bibliography

[1]   Burkard, R. "Sound Pressure Level Measurement and Spectral Analysis of Brief Acoustic Transients." *Electroencephalography and Clinical Neurophysiology* Vol. 57 (1984): 83–91.

[2]   Novak, A., Lotton P., and Simon, L. "Synchronized Swept-Sine: Theory, Application and Implementation." *Journal of the Audio Engineering Society* Vol. 63, No. 10 (2015): 786–798.

12

Discrete Fourier transform

12.1 Discretizing a transfer function

In analyzing a signal on the z-plane up until this point we've only considered sinusoidal signals, which by definition comprise only a single frequency. Music and audio obviously are composed of many frequencies, and therefore we need a tool to analyze the frequency make-up of an arbitrary signal. The z-transform is an obvious place to begin since it is a way to convert a signal or system from the sample domain to a frequency domain.

In Chapter 9, we saw that in the z-domain, the frequency response of a digital filter was found by evaluating the z-transform of that filter on the unit circle. We could take a similar approach to analyze a signal – in Chapter 9.4, the definition of the z-transform for a signal was given in Equation (9.10). Replacing z with $e^{j\omega}$ does indeed give the frequency response, $X(\omega)$ of a signal, $x[n]$.

However, the variable $\omega = 2\pi f / f_s$ is continuous, since the frequency is continuous – this poses a problem for digital systems, which can only represent discrete data. When it came to continuous time-domain signals, the approach was to temporally sample the continuous signal, and replace the variable t (time) with a sample index, n. We will take a similar approach with the continuous frequency variable, ω. If we sample the frequency response, we can replace ω with a discrete frequency index. This process is known as the discrete Fourier transform (DFT).

12.2 Sampling the frequency response

Consider the following first-order IIR filter with difference equation:

$$y[n] = x[n] - 0.25y[n-1] \tag{12.1}$$

The transfer function is

$$H(z) = \frac{Y(z)}{X(z)} = \frac{1}{1 + 0.25z^{-1}} \tag{12.2}$$

And the frequency response is

$$H(\omega) = \frac{1}{\underbrace{1 + 0.25e^{-j\omega}}_{continuous\ function}} \tag{12.3}$$

Since the frequency response is continuous, to discretize it, we should sample it. If we were to capture 8 points along the unit circle, as in Figure 12.1, then equal sub-divisions would be represented as

$$\omega_k = \frac{2\pi k}{N} \tag{12.4a}$$

And

$$f_k = \frac{f_s \cdot k}{N} \tag{12.4b}$$

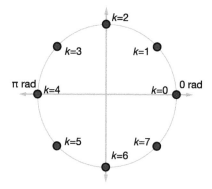

Figure 12.1

The unit circle is sampled at 8 equally spaced frequencies.

Where N is the number of subdivisions (in this example, 8), and k is the sampled frequency index (analogous to n in the sampled time domain).

For this particular example, we have the following ω and f values:

$$\omega_0 = 0, \ \omega_1 = \frac{\pi}{4}, \ \omega_2 = \frac{\pi}{2}, \ ..., \ \omega_7 = \frac{7\pi}{4} \tag{12.5a}$$

$$f_0 = 0, \ f_1 = \frac{f_s}{8}, \ f_2 = \frac{f_s}{4}, \ ..., \ f_7 = \frac{7f_s}{8} \tag{12.5b}$$

If we simply substitute in this sampled ω_k in place of the continuous ω, then the sampled frequency response becomes

$$H[k] = H(\omega)\Big|_{\omega = \omega_k = \frac{2\pi k}{N}} = \frac{1}{1 + 0.25e^{-\frac{j2\pi k}{N}}} \tag{12.6}$$

The discretized frequency response is now given by $H[k]$, using the square brackets to indicate that it is a sampled function.

We can replicate this process for a signal, first starting with the z-transform, then computing the frequency response, then finally sampling the frequency response at equally spaced points around the unit circle. Consider the sequence $x[n]$, with z-transform:

$$X(z) = x[0] + x[1]z^{-1} + x[2]z^{-2} + ... + x[N-1]z^{-(N-1)} \tag{12.7}$$

Then to get the frequency response of this signal, we evaluate on the unit circle:

$$X(\omega) = x[0] + x[1]e^{-j\omega} + x[2]e^{-2j\omega} + ... + x[N-1]e^{-(N-1)j\omega} \tag{12.8}$$

Last, we sample the frequency response at equal points around the unit circle:

$$X[k] = X(\omega)\Big|_{\omega = \frac{2\pi k}{N}} = x[0] + x[1]e^{-1 \cdot \frac{j2\pi k}{N}} + x[2]e^{-2 \cdot \frac{j2\pi k}{N}} ... + x[N-1]e^{-(N-1)\cdot\frac{j2\pi k}{N}} \tag{12.9}$$

12.3 The DFT and inverse discrete Fourier transform

Equation (12.9) can be more compactly written as

$$X[k] \overset{\text{def}}{=} \sum_{n=0}^{N-1} x[n] \cdot e^{-j2\pi \frac{k}{N} n} \tag{12.10}$$

This is the definition of DFT of a digital sequence, $x[n]$, resulting in a sampled frequency response, $X[k]$. Note that $e^{-j2\pi k/N}$ is simply a vector corresponding to one of the sampled locations around the unit circle (determined by k). Then multiplying the exponent by n sets this vector in motion, resulting in a phasor that rotates at different discrete velocities, set by k. In other words, the DFT analyzes the entirety of the sequence x against N complex phasors with different frequencies. If the frequency set by $f_S k/N$, from Equation (12.4b), is present in x, then the magnitude at $X[k]$ will be large, and if that frequency is not present, then the magnitude at that bin will be small.

Importantly, the DFT is completely and transparently invertible – the original sequence $x[n]$ is recoverable from $X[k]$. This process is known as the *inverse discrete Fourier transform* (IDFT), and is given by:

$$x[n] = \frac{1}{N} \sum_{k=0}^{N-1} X[k] \cdot e^{+\frac{j2\pi n}{N}k} \tag{12.11}$$

Note that the equation is very similar to the DFT. The first difference is that the position of x and X are swapped – for the IDFT, every frequency in X contributes to every sample in x. Also, the angle of the phasor is reversed for the IDFT – this simply "undoes" the original multiplication, the inverse of $e^{-anything}$ is $e^{anything}$, and a value times its inverse is always one. The final difference is the scale factor $1/N$. In the DFT, since all samples are being summed, in order to recover the original sample values, X must also be scaled down by a factor of N. Note that it is not uncommon for the $1/N$ scale factor to appear in the DFT equation (and not in the IDFT equation) – the rationale behind this is that the DFT magnitudes will not change their value based on the length (N) of the DFT – simply something to be mindful of.

12.3.1 Example: DFT of a delta

We can predict what the DFT of a delta function, which is simply a digital impulse with a width of a single sample will look like. Impulses are the preferred test stimulus in many audio applications to probe a device under test since they provide equal energy to all frequencies – which is to say, they have a flat magnitude response. The DFT of a delta function is:

$$\Delta[k] = \sum_{n=0}^{N-1} \delta[n] \cdot e^{-\frac{j2\pi k}{N}n}$$

The delta function is non-zero only at $n = 0$, so expanding the sum yields:

$$\Delta[k] = \delta[0] \cdot e^{-\frac{j2\pi k}{N}0}$$

Note that for every k, Δ always evaluates to 1, which is to say the magnitude is the same at every frequency bin.

$$\Delta[0] = 1, \ \Delta[1] = 1, \ ..., \ \Delta[N-1] = 1$$

12. Discrete Fourier transform

12.3.2 Programming example: fast Fourier transform of a delta

The *fast Fourier transform* (FFT) is simply a fast implementation of the DFT that removes redundant calculations, resulting in an efficient algorithm with a computation time on the order of $2N\log(N)$, whereas the DFT is on the order of $2N^2$. The **fft()** function takes two arguments – the input signal, and the number of frequency points to analyze, N. An appropriate size for N is <u>at least</u> as long as the number of samples in the sequence (if and only if perfect reconstruction of the original sequence is desired).

```
N=8;
d = zeros(N,1);
d(1)=1;

plot(abs(fft(d,N)))
grid on; box on;
ylabel('Magnitude'); xlabel('k');
```

The magnitude of the DFT is indeed flat, as expected. Try changing the location of 1 within the vector d – does anything change about the magnitude response? Next, take a look at the angle (use the **angle()** and **unwrap()** functions) – how is the angle affected by delaying the delta?

12.4 Twiddle factor

The DFT contains a complex phasor that analyzes the signal, x. This analysis phasor $\left(e^{-j2\pi\frac{k}{N}n}\right)$ can be separated into two parts: the constant and the variable. Once the N has been selected, then there are only two variables, the frequency bin, k, and the sample index, n – everything else remains constant. This constant, trigonometric value that is being multiplied by the data, is known as the *twiddle factor*, which is represented by W_N, and is given by

$$W_N = e^{-\frac{j2\pi}{N}} \tag{12.12}$$

The twiddle factor represents the number of frequencies between DFT bins – a large N results in a small twiddle factor with fine frequency resolution (and vice versa). The DFT and IDFT can be rewritten using the twiddle factor as

$$X[k] = \sum_{n=0}^{N-1} x[n] \cdot W_N^{kn} \tag{12.13a}$$

$$x[n] = \frac{1}{N} \sum_{k=0}^{N-1} X[k] \cdot W_N^{-kn} \tag{12.13b}$$

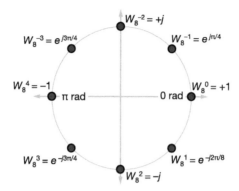

Figure 12.2

Locations around the unit circle for an 8-point DFT.

We can re-label the unit circle in Figure 12.1 using the twiddle factor for $N = 8$, as shown in Figure 12.2. Notice that increasing orders of W_8 move in the clockwise direction around the unit circle – this is because in Equation (12.12) the twiddle factor includes a negative sign in the exponent.

12.4.1 Programming example: DFT of a sinusoid

When we look at the argument of a digital sinusoid, we expect to see $(2\pi f n/f_S)$, where the fraction f/f_S gives us a value between 0 and 0.5, placing the frequency on the unit circle somewhere between 0 and π rad. Here, let's consider a digital sequence with length N that is given by

$$x[n] = \cos\left(\frac{2\pi r n}{N}\right)$$

Therefore, the frequency must be the ratio r/N. Importantly, in this example, this puts the frequency of the cosine exactly equal to one of the analysis frequencies, k, when we perform the DFT on x (if r is an integer). The cosine can be rewritten as a summation of two complex conjugates, each with half the magnitude:

$$x[n] = \frac{1}{2}\left(e^{j\frac{2\pi r n}{N}} + e^{-j\frac{2\pi r n}{N}}\right)$$

Even though the DFT has not been applied yet, nothing is preventing us from applying Equation (12.12) to x, which can be rewritten using W_N as

$$x[n] = \frac{1}{2}\left(W_N^{-rn} + W_N^{rn}\right)$$

Even though it looks a bit unconventional, this is still the same N-point cosine. Next, we can perform the DFT and apply the distributive property, yielding

12. Discrete Fourier transform

$$X[k] = \frac{1}{2}\sum_{n=0}^{N-1} W_N^{-rn} \cdot W_N^{kn} + \frac{1}{2}\sum_{n=0}^{N-1} W_N^{rn} \cdot W_N^{kn}$$

After some algebra:

$$X[k] = \frac{1}{2}\sum_{n=0}^{N-1} W_N^{(k-r)n} + \frac{1}{2}\sum_{n=0}^{N-1} W_N^{(k+r)n}$$

When $k \neq r$, then the sum is 0. Only when the exponent is 0, then the sum is $N/2$, yielding

$$X[k] = \begin{cases} \dfrac{N}{2}, & k=r, k=-r \\ 0, & \textit{otherwise} \end{cases}$$

$X[k]$ has a large magnitude at the two frequency bins associated with r. If we imagine this cosine on the z-plane, it comprises two phasors at $\pm r/N$ – conjugates, each with half the amplitude, as in Figure 12.3(a). If we were to imagine the unit circle "un-rolled", then we would see frequency on a horizontal axis, as in Figure 12.3(b).

This result is easy to visualize in MATLAB®/Octave – for no particular reason, let's select a $N = 32$ and $r = 4$ (can you determine the frequency of this cosine for f_S = 48 kHz?). We can verify that $\frac{1}{2}\left(W_N^{-rn} + W_N^{rn}\right)$ is indeed a cosine; however, due to tiny numerical errors, the imaginary components don't completely cancel, leaving

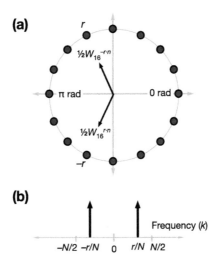

Figure 12.3
(a) A cosine comprises two complex conjugate phasors. (b) The magnitude of these phasors is shown as a function of frequency.

a residue that is smaller than the LSB of a 32-bit signal. For this reason, the signal cos_r is technically complex (in practice only – mathematically it is purely real), and so the **real()** function must be applied for plotting purposes. The DFT is implemented as nested for-loops, and the output is printed and plotted, with DC at $k = 0$ and Nyquist at $k = 16$. The magnitude response resembles Figure 12.3(b).

```
N=32;
r=4;
W=exp(-j*2*pi/N); % twiddle factor
cos_r = 0.5*W.^(r*[0:N-1]) + 0.5*W.^(-r*[0:N-1]);
plot(real(cos_r))

for k=0:N-1

    X(k+1)=0;
    for n=0:N-1
        X(k+1)=X(k+1)+0.5*W^((k-r)*n)+0.5*W^((k+r)*n);
    end
    disp(['|X(k=' num2str(k) ')| = ' num2str(abs(X(k+1)))]);

end

figure; plot(0:N-1,(abs(X)))
```

12.4.2 Example: DFT of a rectangular pulse

In Chapter 4.4, we saw that an ideal low-pass filter, defined with a rectangular pulse in the frequency domain, corresponded to a sinc function in the time domain. In fact, this relationship works in both directions – a rectangular pulse in the time domain results in a sinc shape in the frequency domain. We can observe this by performing the DFT on the following signal:

$$x[n] = \begin{cases} 1, & 0 \leq n < 4 \\ 0, & 4 \leq n < 8 \end{cases}$$

To find the DFT, we must first recognize that $N = 8$, therefore the possible values produced by the twiddle factor are shown in Figure 12.2. Next, when applying the DFT, we see that since $x[n]$ is 0 for samples 4 through 7, then the computation of the DFT is simplified somewhat.

$$X[k] = \sum_{n=0}^{7} x[n] \cdot W_8^{nk}$$

$$= \sum_{n=0}^{3} 1 \cdot W_8^{nk} + \sum_{n=4}^{7} 0 \cdot W_8^{nk}$$

Expanding the summation, the DFT can be written as follows:

$$X[k] = W_8^0 + W_8^k + W_8^{2k} + W_8^{3k}$$

Starting at $k = 0$, since the signal x has a positive and non-zero mean value, then we expect a non-zero value at the DC bin:

$$X[0] = W_8^0 + W_8^{1 \cdot 0} + W_8^{2 \cdot 0} + W_8^{3 \cdot 0}$$

$$= 1 + 1 + 1 + 1 = 4$$

At $k = 1$, we refer back to Figure 12.2 to find the values for each order of W_8, resulting in a complex value. Notice that the order of W_8 increases by one each subsequent term:

$$X[1] = W_8^0 + W_8^{1 \cdot 1} + W_8^{2 \cdot 1} + W_8^{3 \cdot 1}$$

$$= 1 + e^{-\frac{j\pi}{4}} - j + e^{-\frac{j3\pi}{4}}$$

$$= 1 - j2.414 = 1.848 \angle -\frac{3\pi}{8}$$

Then at the next frequency bin, $k = 2$, the orders increase by 2 each term:

$$X[2] = W_8^0 + W_8^{1 \cdot 2} + W_8^{2 \cdot 2} + W_8^{3 \cdot 2}$$

$$= 1 - j - 1 + j$$

$$= 0$$

When $k = 3$, the final term W_8^9 is at the same location on the unit circle as W_8^1:

$$X[3] = W_8^0 + W_8^3 + W_8^6 + \overbrace{W_8^9}^{W_8^1}$$

$$= 1 + e^{-\frac{j6\pi}{8}} + j + e^{-\frac{j2\pi}{8}}$$

$$= 1 - j0.414 = 1.189 \angle -\frac{\pi}{8}$$

Finally, at the Nyquist bin, $k = 4$, we see that increasing W_8 by orders of four each term simply results in alternating between DC and Nyquist on the unit circle:

$$X[4] = W_8^0 + W_8^4 + \overbrace{W_8^8}^{W_8^0} + \overbrace{W_8^{12}}^{W_8^4}$$

$$= 1 - 1 + 1 - 1$$

$$= 0$$

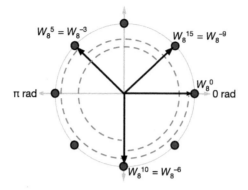

Figure 12.4

W_8^0, W_8^5, W_8^{10}, and W_8^{15} correspond with W_8^0, W_8^{-3}, W_8^{-6}, and W_8^{-9}, respectively.

Starting at $k = 5$, increasing W_8 by orders of 5 clockwise around the unit circle is identical to increasing by orders of 3, but in the counter-clockwise direction, as shown in Figure 12.4.

In other words, W_8^{5n} is a conjugate of W_8^{3n}:

$$X[5] = W_8^0 + W_8^5 + \overbrace{W_8^{10}}^{W_8^2} + \overbrace{W_8^{15}}^{W_8^7}$$

$$= 1 + e^{-\frac{j10\pi}{8}} - j + e^{-\frac{j14\pi}{8}}$$

$$= 1 + j0.414 = 1.189\angle\frac{\pi}{8}$$

$$= X^*[3]$$

Predictably, the same phenomenon occurs, whereby $X[6] = X^*[2]$ and $X[7] = X^*[1]$. Plotting the magnitude of $X[k]$, the absolute value of a sinc pattern emerges, in which the main lobe appears centered at $k = 0$ and $k = 8$, as shown in Figure 12.5.

12.5 Properties of the DFT

There are four variants of Fourier analysis. For continuous (i.e., analog) signals, there are two types: the *Continuous Fourier Transform* for aperiodic signals (e.g., a rectangular pulse) that results in a continuous frequency transform, and the *Fourier Series* for periodic signals (e.g., a square wave) that results in a frequency transform with energy only at discrete frequencies (i.e., harmonics). For digital signals, we already saw the two types: the *Discrete-time Fourier Transform* for digital signals with a continuous frequency transform, and the DFT for digital signals, resulting in a discrete frequency transform.

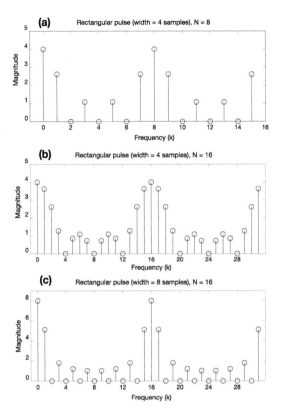

Figure 12.5

(a) The magnitude of the DFT of the rectangular pulse (pulse width of four samples) worked out in the example above is shown for length $N = 8$. (b) If the pulse width is fixed and N is increased to 16, then the overall shape of the magnitude response remains the same, but with greater frequency resolution. (c) If the pulse width is doubled to a width of 8, and N is kept at 16, then the frequency resolution stays the same as (b), but more of the energy moves to 0 Hz, and the sinc main lobe width decreases.

Among these signals and their corresponding spectra, there is a pattern in which an aperiodicity in one domain demands continuity in the other domain, while periodicity in one domain demands the other domain to be discrete.

Signal processing on a computer requires that both the time and frequency domain must be discrete. But in order to accomplish this, both domains must also be periodic. In the frequency domain, this makes sense – the spectrum repeats itself every f_S Hz. However, in the time domain, no such periodicity is guaranteed, or even to be expected. Therefore, for the DFT to be mathematically valid, we must assume that the entire sequence $x[n]$ is repeated forever in each direction. This is known as the **periodicity** property of the DFT, in which

$$x[n+N] = x[n] \tag{12.14a}$$

And

$$X[k+N] = X[k] \qquad (12.14b)$$

Since both $x[n]$ and its DFT, $X[k]$, are both periodic with period N, the following time-reversal property applies:

$$x[-n] = x[N-n] \overset{DFT}{\Leftrightarrow} X[-k] = X[N-k] \qquad (12.15)$$

Equation (12.15) tells us that a time-reversal of a sequence results in the spectrum being mirrored about D.C., or put differently, the conjugate is applied to the entire spectrum, thus flipping the sign on the phase of the spectrum.

Like transfer functions and convolution, the DFT is **linear, commutative**, and **associative**, and the **time-shift** and **convolution** theorems apply. These are described in Chapter 9 by Equations (9.16) through (9.20).

12.6 Revisiting sampling in the frequency domain

If you recall Chapter 3, the process of sampling involved multiplying a continuous function, $x_c(t)$, by a delta train, $d(t)$, that had spacing of T_S between delta locations. The Fourier transform of a delta train is given by

$$D(f) = F_S \cdot \sum_{k=-\infty}^{\infty} \delta(f - k \cdot f_S) \qquad (12.16)$$

The delta train, which is discrete and periodic, has a frequency transform, $D(f)$, that is also discrete and periodic. $D(f)$ is also a delta train with spacing of f_S Hz, since $f-kf_S$ is always 0 except when f is an integer multiple of f_S, as shown in Figure 12.6.

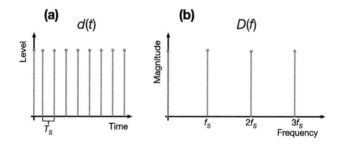

Figure 12.6

(a) A delta train with delta spacing of T_S. (b) The DFT of a delta train results in a delta spacing of $1/T_S$, or f_S.

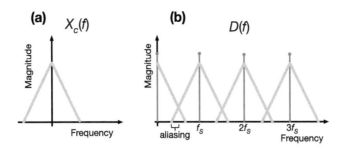

Figure 12.7

(a) The spectrum of a continuous signal is shown. (b) When sampled, the spectrum repeats itself every f_S Hz. If the continuous spectrum has any magnitude at frequencies greater than f_N, then aliasing (overlap of the spectra) will occur.

Recall that a sampled signal, $x[n]$, is a continuous signal multiplied by a delta train:

$$x[n] = x_c(t) \cdot d(t) \tag{12.17}$$

If we let the Fourier transform of $x_c(t)$ be $X_c(f)$, then Equation (12.17) can be equivalently written in the frequency domain as

$$X(f) = X_c(f) * D(f) \tag{12.18}$$

Note that multiplication in the time domain becomes convolution in the frequency domain. What Equation (12.18) is telling us is that in order to obtain the spectrum of a sampled signal, we must convolve the spectrum of the continuous signal by the spectrum of a delta train. The result of convolving the spectrum in Figure 12.6(b) with the spectrum in Figure 12.7(a) is the spectrum shown in 12.7(b).

Note that aliasing in Figure 12.7(b) appears where the spectrum of $X_c(f)$ extends beyond $f_S/2$. Aliasing occurs when the continuous signal is undersampled, or not low-pass filtered below the Nyquist frequency.

12.7 Frequency interpolation

In Example 12.4.1 we considered the DFT of a sinusoid in which the frequency of the sinusoid was exactly the frequency of one of the DFT bins. While there is no singly, typical DFT length, it is quite common in audio applications (especially real-time processes) to see an N in the range of 64 up to 2048. Let's consider $N = 1024$ and a sample rate of $f_S = 44{,}100$ Hz. For this particular configuration, we expect a frequency spacing of

$$\Delta f = \frac{f_S}{N} = \frac{44{,}100}{1{,}024} = 43.06640625 \; Hz \tag{12.19}$$

Therefore, the frequencies exactly represented by a DFT bin are listed in Table 12.1.

Table 12.1 Notes Corresponding to the First Seven Bins of a
1024-Point DFT at a Sample Rate of 44.1 kHz

k	f (Hz)	Note (re: A4 = 440 Hz)
0	0	n.d.
1	43.1	F1 (−23.5 cents)
2	86.1	F2 (−23.5 cents)
3	129.2	C3 (−21.5 cents)
4	172.3	F3 (−23.5 cents)
5	215.3	A3 (−37 cents)
6	258.4	C4 (−21.5 cents)

Two things should immediately be apparent regarding this frequency spacing. First, especially at low frequencies, the bin spacing is very coarse compared to a semi-tone. For example, bins 1 to 4 cover two octaves! Clearly, this poor frequency resolution misses a lot of notes. But also, the bin frequencies don't even fall on notes from an A4 = 440 Hz, equal-temperament scale. For these reasons, we can expect that most musical notes will actually exist between DFT bins.

This raises the question: what happens to the DFT when the signal frequency falls between two analysis frequencies? The answer is that the magnitude becomes split between the two most adjacent bins, and the closer bin gets the majority of the magnitude. The "true" underlying frequency can be estimated by considering the two adjacent bins, one higher and one lower, via interpolation. The frequency can be estimated by

$$k_d = \frac{\left(\left|X(k+1)\right| - \left|X(k-1)\right|\right)}{\left(\left|X(k+1)\right| + \left|X(k-1)\right|\right)} \qquad (12.20a)$$

$$f_{est} = (k + k_d) \cdot \frac{f_s}{N} \qquad (12.20b)$$

12.7.1 Programming example: frequency interpolation

Consider a 1024-point sinusoid sampled at 44.1 kHz. The frequency spacing is already given by Equation (12.19) and Table 12-1. Let's consider two frequencies in particular – the first, x_1, will be forced to the 24th bin, given as $24f_S/N$, or approximately 1.0336 kHz. The second sinusoid, x_2, will have a frequency of 1.0000 kHz, which falls between the 23rd bin (about 0.9905 kHz) and the 24th bin. The frequency interpolation algorithm estimates the frequency of x_2 to be at 1.0003 kHz – only 0.3 Hz off from the true frequency, but much better than the 9.5 Hz error without using frequency interpolation (Figure 12.8).

12. Discrete Fourier transform

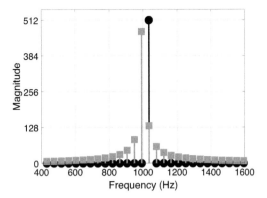

Figure 12.8

The magnitudes of X_1 and X_2 are shown for a 1024-point DFT. The frequency of x_1 (black, circle) lands exactly on a DFT bin, while the frequency of x_2 (gray, square) lands between two bins. Since the frequency in x_1 is closer to the 23rd bin (0.9905 kHz), then the majority of the magnitude appears here.

```
N=1024;
n=0:N-1;
fs=44100;
k=24;


x1=sin(2*pi*(k/N*fs)*n/fs);
x2=sin(2*pi*1000*n/fs);

X1=fft(x1, N);
X2=fft(x2, N);

f=linspace(0,fs,N);

stem(f,abs(X1));
hold on;
stem(f,abs(X2));

yp1=abs(X2(k+1));
ym1=abs(X2(k-1));
d = (yp1 - ym1)/(yp1+ym1);
f_est = (k+d-1)*fs/N
```

12.8 Challenges

1. Consider the following digital sequence:

$$x[n] = 0.1 + 0.2\cos\left(\frac{2\pi n}{16}\right) + 0.2\cos\left(\frac{2\pi n}{32}\right)$$

Sketch the magnitude response of this signal for $n = [0, …, 127]$. Label frequency bins (horizontal axis) and magnitude levels (vertical axis) for each frequency component. *Hint: you should have 5 spectral components.*

2. Write a MATLAB®/Octave function for computing the DFT. This function should contain two nested for-loops: one for iterating over the frequency bins, k, and one for iterating over the sample indexes, n.
   a. Create a 1024-point sequence and compute the DFT using your function.
   b. Use the tic and toc commands to measure how long the function takes to execute.
   c. Next, measure how long the fft command takes on the same sequence. How do these compare?

3. An audio track has a sample rate of 48 kHz. It is analyzed in 10 ms windows with the DFT.
   a. What is the frequency resolution (in other words, how many Hz between frequency bins)?
   b. Which bin(s) of the spectrum $X[k]$, $k = [0, 1, …, N–1]$ contain the frequency 120 Hz?

4. Consider the following digital sequence:

$$x[n] = [-1, 1, -1, 1]$$

   a. Compute $X[k] = DFT\{x[n]\}$
   b. Perform the IDFT on $X[k]$ to demonstrate perfect reconstruction from time to frequency domain and back.

12.9 Project – spectral filtering

The FFT is an efficient algorithm for identically computing the DFT of a sampled signal in a fraction of the time. Many FFT algorithms exist, and their exact function is outside the scope of this text; however, the basic premise is that repeated calculations of the twiddle factor can be avoided by taking advantage of mathematical symmetries around the unit circle.

Just like the DFT, the FFT produces the same number of output samples as input samples, has frequency bins from 0 Hz to f_S Hz, spaced to f_S/N Hz apart, and the first bin is 0 Hz. However, unlike the DFT, the FFT is much faster to compute when the input length is a power of 2 (for example, 128, 256, 1024, etc.).

Importantly, when transforming to the frequency domain, a convolution operator in the sample domain becomes a multiplication operator in the frequency domain. Therefore, rather than obtaining an output of a filter via convolution, it is often computationally cheaper to transform a signal to the frequency domain (with the FFT), perform filtering operations using element-wise multiplication, followed by transformation back to the sample domain with the inverse fast Fourier transform (IFFT). This might seem unbelievable at first, but consider a 1024-point FIR filter and a 1024 length sequence.

For convolution, the number of mathematical operations (multiplies and adds) are $2 \cdot 1024 \cdot 1024 = 2{,}097{,}152$ operations. On the other hand, the FFT takes $2 \cdot 1024 \cdot \log_{10}(1024)$

= 6,165 operations. There will, of course, be another 6,165 operations for the IFFT, and as well 1,024 multiplications (of magnitudes) and 1,024 summations (of phases) to perform frequency domain filtering, for a grand total of 14,378 operations. This represents a computational savings of over 99%.

In this lab, in order to measure the amount of time required to perform various algorithms, we will make use of the commands **tic** and **toc**. These are timing commands that start a timer (**tic**) and stop a timer (**toc**). Additionally, the **toc** command can return an output for saving the elapsed time in a variable.

Assignment

1. Load in a few seconds of audio from any source, and convert to mono.
2. Write a convolution function, following Programming example 8.3.3.
3. Write a frequency domain convolution function that takes two signals, transforms them to the frequency domain using FFT, carries out the filtering operation in the frequency domain, and then transforms back to the sample domain using IFFT.
4. With two test signals, verify that problems 2 and 3 produce identical outputs to the **conv**() function.
5. Convolve a snare hit with a room IR (both available from DigitalAudioTheory.com, or simply use any IR that you have at your disposal) using both functions you wrote in Problems 2 and 3. Use the **tic** and **toc** commands to measure how long each function takes.
6. Calculate the computational savings of transforming to the frequency domain to perform convolution. Are there any disadvantages?

13

Real-time spectral processing

As you hopefully saw in the Chapter 12 Project, processing a digital signal through a digital system (such as a filter) can be much more efficient in the frequency domain, a practice known as *spectral processing*. In addition to increasing processing speed, spectral processing also opens the door for designing an arbitrary magnitude response for a filter, giving specific control over filter design. Spectral processing can also be used for advanced signal analysis, allowing for the visualization of a spectrum over time, known as a *spectrogram*. These are popular in many audio repair/restoration and spectral filtering plugins.

However, there is a trade-off when transforming to the frequency domain – to increase frequency resolution, we must increase the number of input samples, thereby decreasing temporal resolution. This is because all the samples being transformed are being considered together. If a filter is expected to change over time (for example, in the case of automation), then we cannot simply process an entire audio track all at once. The solution to this is to *frame* the audio into segments, typically powers of two to maximize FFT efficiency. Typical frame sizes in a DAW are 128, 256, 512, etc. – increasing frame size also increases playback latency (calculated as the frame length divided by the sample rate).

13.1 Filtering in the frequency domain

Let's look at filtering in the frequency domain. Recall that while filtering is achieved by convolution in the sample domain, in the frequency domain this becomes element-by-element multiplication. This implies that the lengths of the signal and the filter ought to be the same. This is not necessarily true in the generic case – the signal length will be equal to the frame size, which may often be set by the DAW/engineer, while the filter length is arbitrarily long, and is set by the plug-in itself. This discrepancy is easy to overcome by increasing N during the frequency transformation to match the length of whichever signal is longer. Furthermore, N will typically be increased to the next power of 2, to increase FFT efficiency.

Let's borrow the signal from Example 12.4.2 in the previous Chapter, which was a rectangular pulse, where $x[n] = [1, 1, 1, 1, 0, 0, 0, 0]$, and the frequency transform was $X[k] = [4\angle 0, 1.85\angle -3\pi/8, 0, 1.19\angle -\pi/8, 0, 1.19\angle \pi/8, 0, 1.85\angle 3\pi/8]$. This signal will be filtered with an FIR filter given by $h[n] = [0.5, 0, 0.5]$. As a first step, let's take the z-transform of $h[n]$, and replace z with ω_k, such that $H[k] = 0.5 + 0.5e^{-2j2\pi k/N}$. The original length of $h[n]$ was 3, so N must be at least that long, but since our signal $x[n]$ has a length 8, we will select $N = 8$ to match. Solving for $H[k]$:

$$H[0] = 0.5 + 0.5W_8^0 = 1\angle 0 \tag{13.1a}$$

$$H[1] = 0.5 + 0.5W_8^2 = 0.5\angle -\frac{\pi}{4} \tag{13.1b}$$

$$H[2] = 0.5 + 0.5W_8^4 = 0\angle 0 \tag{13.1c}$$

$$H[3] = 0.5 + 0.5W_8^6 = 0.5\angle \frac{\pi}{4} \tag{13.1d}$$

$$H[4] = 0.5 + 0.5W_8^8 = 1.0\angle 0 \tag{13.1e}$$

$$H[5] = H^*[3] = 0.5\angle -\frac{\pi}{4} \tag{13.1f}$$

$$H[6] = H^*[2] = 0\angle 0 \tag{13.1g}$$

$$H[7] = H^*[1] = 0.5\angle \frac{\pi}{4} \tag{13.1h}$$

It can be seen that $H[k]$ is a comb filter, which could have been predicted from the IR, which shows a signal being added to a delayed version of itself. Now the output, $Y[k]$, is given by

$$Y[k] = X[k] \cdot H[k] \qquad (13.2)$$

The output is computed by multiplying together the magnitudes and summing the phases of X and H of corresponding bins, k, whereby:

$$X[k] = \left[4\angle 0 \quad 1.85\angle\left(-\frac{3\pi}{8}\right) \quad 0 \quad 1.19\angle\left(-\frac{\pi}{8}\right) \quad 0 \quad 1.19\angle\left(\frac{\pi}{8}\right) \quad 0 \quad 1.19\angle\left(\frac{\pi}{8}\right) \right]$$

$$H[k] = \left[1\angle 0 \quad 0.5\angle\left(-\frac{\pi}{4}\right) \quad 0 \quad 0.5\angle\left(\frac{\pi}{8}\right) \quad 1 \quad 0.5\angle\left(-\frac{\pi}{4}\right) \quad 0 \quad 0.5\angle\left(\frac{\pi}{4}\right) \right]$$

$$Y[k] = \left[4\angle 0 \quad 0.925\angle\left(-\frac{5\pi}{8}\right) \quad 0 \quad 0.595\angle\left(\frac{\pi}{8}\right) \quad 0 \quad 0.595\angle\left(-\frac{\pi}{8}\right) \quad 0 \quad 0.925\angle\left(\frac{5\pi}{8}\right) \right]$$

$$(13.3)$$

Note that Y from $k = 5$ through $k = 7$ is conjugate with Y from $k = 3$ through $k = 1$, which means that $y[n]$ will be purely real (a necessary requirement for audio, and the expected outcome for convolving two real signals). The final step is to transform back to the sample domain, using the IDFT.

$$y[n] = \frac{1}{8} \sum_{k=0}^{7} Y[k] \cdot W_8^{-nk} \qquad (13.4a)$$

$$y[n] = \frac{1}{8} \left[\begin{array}{l} Y[0] \cdot W_8^0 + Y[1] \cdot W_8^{-n} + Y[2] \cdot W_8^{-2n} + Y[3] \cdot W_8^{-3n} + \\ Y[4] \cdot W_8^{-4n} + Y[5] \cdot W_8^{-5n} + Y[6] \cdot W_8^{-6n} + Y[7] \cdot W_8^{-7n} \end{array} \right], n = 0\ldots7 \qquad (13.4b)$$

$$y[n] = [0.5, 0.5, 1.0, 1.0, 0.5, 0.5, 0, 0] \qquad (13.4c)$$

This example demonstrates the general approach to spectral processing involving a filter and a signal. While an FIR filter was used here, the process using an IIR filter is no different.

13.2 Windowing

In the above example, we considered the entirety of the input signal, x. However, as previously mentioned, in audio processing, we are normally framing a much longer signal into shorter chunks. This process can be thought of as multiplying the signal by an envelope. In this case, the envelope is a rectangular pulse (often referred to as a boxcar), as shown in Figure 13.1. Spectral processing is performed on this short frame, with the results saved to an output buffer. The boxcar is then moved to a later

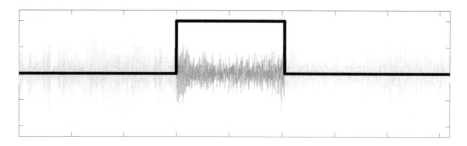

Figure 13.1

An audio signal (gray) is framed by a boxcar (black) prior to spectral processing.

moment in time, picking up with the very next sample from where the previous frame left off, and spectral processing is performed once again on the next frame.

Since the process of framing involves multiplying an envelope (the boxcar) by a signal in the sample domain, this becomes equivalent to convolving their respective spectra in the frequency domain. In other words, the spectrum of the boxcar <u>interferes with</u> the spectrum of the audio that it is framing out, which is unfortunate. This raises the question: What is the spectrum of a boxcar?

As we saw in Chapter 4.4, the frequency transform of a rectangular pulse, or boxcar, is a sinc function, whose main lobe width is inversely proportional to the boxcar width. Convolving the sinc with the spectrum of the audio has the effect of spreading out the energy. The main lobe spreads out the energy around each spectral line, while the side lobes also introduce cross-talk at remote frequencies, as shown in Figure 13.2. Longer frame sizes result in sincs with narrower main lobes and lower sidebands.

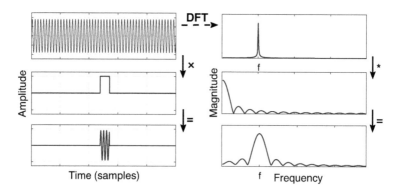

Figure 13.2

A sample domain sinusoid with frequency *f* (top-left) and its magnitude response (top-right). A boxcar window and magnitude response (middle-left and -right). The framed signal is a result of multiplying the signal by the boxcar (bottom-left). The resulting magnitude response is a convolution of the above two magnitude responses (bottom-right).

Another problem that can occur when framing audio is when a frequency falls between two DFT bins. When this happens, the energy gets divided between the two most adjacent bins. This creates an apparent (although not real) drop in level for these "in-between" frequencies, a phenomenon known as *scalloping*. Boxcar windows exhibit especially bad scalloping with magnitude discrepancies up to 3.9 dB between a frequency that falls exactly on a bin and one that falls half-way between two bins.

The problems of cross-talk (related to the level of side lobes) and scalloping can both be solved by more smoothly tapering the shape from a boxcar to a function that transitions from 0 (or thereabouts) to 1 and back to 0 again. These shapes are governed by equations known as *window* functions (or simply "windows"), and the process of multiplying a framed signal by a window is known as windowing.

Selecting a window is a balancing act between various window properties – there is no "perfect" window, only some that are better suited for certain applications (see Figures 13.3 and 13.4). Narrow main lobe windows have the best frequency resolution, but they also exhibit greater scalloping and cross-talk. High-resolution windows perform well in high noise scenarios. On the other hand, wide main lobe windows have a very good representation of signal level, but with poorer frequency resolution. Windows can also be designed to have very low cross-talk (for example, the Blackman-Harris window), which reduces the noise floor, thus making these optimal for high dynamic range scenarios. In the absence of application-specific information, the Hann window is a good default window for audio – it has minimal scalloping, decent frequency resolution, and sidebands that roll-off at –18 dB/octave. Additionally, the Hann window has attractive temporal properties that will be examined in the next section.

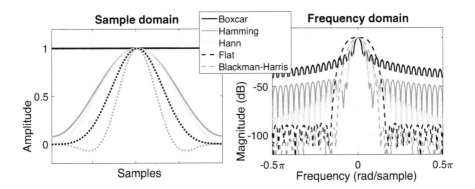

Figure 13.3

Several windows are shown here from those with very good frequency resolution (boxcar) to those with very accurate level capture (flat). The Hamming and Hann are good compromise windows.

Some notable characteristics of commonly used audio windows are listed below:

- Hann window: most commonly used when an inverse transformation back to the sample domain is required; side lobes adjacent to the main lobe are –31 dB but have a steep roll-off (–18 dB/octave); a good all-around window
- Hamming window: first side lobes are completely cancelled out, and the next side lobes are lower than the Hann window at –41 dB, but with only a –6 dB/octave roll-off; good for frequency analysis
- Tukey window: adjustable window shape; good when window size may change, as transition windows are possible

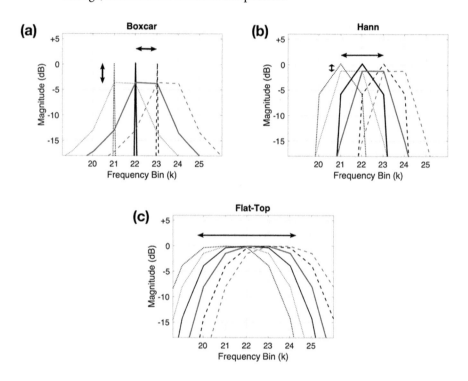

Figure 13.4

Three windows are shown, each with six sinusoids. Three sinusoids have a frequency that falls directly on a bin (black solid, dashed, and dotted), while the other three have frequencies that are half-way between two bins (gray solid, dashed, and dotted). Dashes are just to help with visualization and don't indicate anything. Scalloping is indicated by the vertical arrows, while lobe width by the horizontal arrows. (a) The boxcar window has the best frequency resolution and level accuracy of all windows, but only for frequencies landing directly on a bin. They also exhibit the most scalloping (3.9 dB), and the frequency resolution varies depending on where the frequency lands. (b) A Hann window has less scalloping (1.4 dB) and has a frequency resolution that is twice as wide as the boxcar and with more even lobe widths between on- and off-bin frequencies. (c) The flat-top window exhibits absolutely no scalloping but has the worst frequency resolution (2.5 times as wide as the Hann window).

13. Real-time spectral processing

- Blackman-Harris window: lowest sidebands at −92 dB; main lobe width that is twice as wide as Hann and Hamming windows
- Chebyshev window: selectable sideband level; sidebands do not roll-off away from the main lobe; lowering the sideband level widens the main lobe width
- Flat-top window: exhibits no magnitude scalloping; sidebands that are −83 dB; poor frequency resolution

13.3 Constant overlap and add

One problem that is probably already obvious with windowing a frame of audio is that the levels are being manipulated, resulting in a pulsating envelope. Even if the audio is only being analyzed (and not reconstructed in the sample domain), the process of windowing weights the samples near the middle of the window more than the ones near the edge, potentially missing important audio information. The solution to this is to design windows such that they can be overlapped, and when the overlapped regions are summed together, they produce a constant value of 1, a technique known as *constant overlap and add* (or COLA). The amount of overlap is usually specified by the proportion of the overlap amount to the total window length. Shown in Figure 13.5 is a series of overlapping Hann windows and their sum, which is equal to 1.

Note that for COLA windows, when the value of one window reaches 1, the values of any of the adjacent windows is exactly 0 at this same sample index. For 50% overlapped windows, the overall efficiency of spectral processing is reduced by a factor of 2 since every sample is actually processed twice. Also, note that two half-length windows are required for the first and last windows to preserve COLA.

13.4 Spectrograms

Spectrograms are the graphical representation of the frequencies in a signal as they evolve over time. Usually, COLA windows are utilized in a spectrogram plot, with time on the horizontal axis, frequency on the vertical axis, and magnitude encoded in color or intensity. Decreasing the length of the window gives better temporal resolution, while increasing the length of the window gives better frequency resolution. One

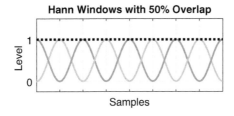

Hann Windows with 50% Overlap

Figure 13.5

A series of 50% overlapping Hann windows are shown (alternating gray, solid) along with the sum of the windows (black, dashed).

way to gain both temporal and frequency resolution (it shouldn't be possible, right?) is to employ a window that tolerates more than 50% overlap. For example, the Hamming window supports 75% overlap (each sample is covered by four windows) while the Blackman-Harris supports 66.6% overlap (three windows per sample).

13.4.1 Programming example: vocal formants

Spectrograms are an interesting way of visualizing speech, since the harmonics produced during speech production (known as formants) are known to have levels that change depending on the phoneme that is being spoken. In this example, you will first record three seconds of your voice producing the phoneme /æ/ (the vowel sound in "cat"), then you will visualize the formants with a spectrogram, and finally isolate a single formant to band-pass filter using real-time spectral processing.

The first step is to record a vocal sound using the **audiorecorder**(), and to extract this data into a variable, x. It may be helpful to use the higher range of your voice to spread the formants out a bit. We will truncate the length to a power of 2. When you run this section of code, the microphone will turn on and perform an audio capture – so be ready!

```
% capture audio
Fs=44100;
nbits=16;
nchans=1;

vox=audiorecorder(Fs, nbits, nchans);
recordblocking(vox, 3);

play(vox);

x=getaudiodata(vox,'double');
N=2^17;
x=x(1:N);
```

Next, we will visualize the phoneme using **spectrogram**(). This function analyzes x, with user-specified window length (here, 1024 samples), overlap amount (93.75%), sample rate, and the preferred location for the frequency axis (vertical, or y-axis). You should have a figure that resembles Figure 13.6. Notice the presence of several harmonics above the fundamental pitch. I will select the 7th harmonic, which happens to be the $\flat 7$ relative to the fundamental, to isolate and band-pass filter. For my voice, this is around 1400 Hz, so I will create a fourth-order BPF using **butter**() centered around this frequency location, and convert it to the frequency domain for spectral processing using **freqz**(). The argument 'whole' needs to be added to the **freqz**() command so that the entire spectrum up to f_S is returned.

```
% spectrogram
NFFT=1024;
overlap=NFFT/16;
figure()
% for Matlab use:
spectrogram(x, NFFT, overlap, NFFT, Fs, 'yaxis');
% for Octave use:
specgram(x, NFFT, overlap);
```

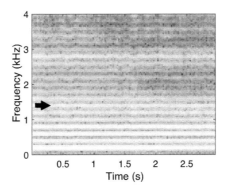

Figure 13.6

A spectrogram of my voice producing the phoneme /æ/. The 7th harmonic is indicated with an arrow.

```
% define filter
Fn=Fs/2;        %nyquist
ord=4;          %order
fl=1250;        %hz
fh=1550;
[a,b]=butter(ord, [fl/Fn, fh/Fn]);
[H,~]=freqz(a,b,NFFT, 'whole');
```

Next, the preprocessing steps will be performed. This first involves creating the windows; here a Hann window of a length of 1024 samples (corresponding to about 23.2 ms) will be used. Additionally, special half windows need to be created for the first and last 512 samples of the signal. The last preprocessing step is to define the "jump" size and determine the number of frames. With 50% overlap, we know there will be twice as many frames as there would be with no overlap (minus one, since we lose half a window at the onset and offset). And the jump size, or distance from the start of one window to the start of the next, is simply the overlap amount times the window length.

```
% overlap and add
hannWin=hann(NFFT);

% make special first window and last window
firstWin = hannWin;
firstWin(1:end/2)=1;
lastWin=hannWin;
lastWin(end/2+1:end)=1;

% define output to match input size
y=zeros(size(x));

% how many frames?
numFrames = 2*length(x)/NFFT-1;

jump = NFFT/2;
```

Finally, we move on to the actual processing. We iterate over all the frames and first select the window to use, followed by calculation of the start and end sample indexes of the window. Next, a frame of audio is extracted according to these indexes and windowed, then converted to the frequency domain.

At this point, we can do our spectral multiplication with the BPF, and convert the product back to the sample domain with **ifft**(). Note the use of the **real**() function – even though the imaginary portions are fully cancelled mathematically, sometimes computationally a tiny residue remains, which should be ignored and discarded.

```
for ii=1:numFrames
    % window selection
    if ii==1
        win=firstWin;
    elseif ii==numFrames
        win=lastWin;
    else
        win=hannWin;
    end

    %frame and window audio
    startIdx = (ii-1)*jump+1;
    endIdx   = startIdx+NFFT-1;

    frameBox = x(startIdx:endIdx);
    frame = frameBox.*win;

    % do spectral processing
    X=fft(frame,NFFT);
    FRAME=X.*H;
    frame=real(ifft(FRAME,NFFT));

    % overlap and add
    y(startIdx:endIdx) = y(startIdx:endIdx) + frame;
end

sound(y, Fs);
```

You can repeat this final section of code with different BPF center frequencies to hear each of your formants. Or contrarily, you can switch instead to a band-stop filter to remove certain formants and hear how your voice changes.

13.5 Challenges

1. Using spectral processing, convolve two rectangular pulses. Make each signal 256 samples in length, with a pulse width of 64 samples. You should obtain a triangular shape.

2. What is the impact on temporal and frequency resolution when a window length is decreased?
3. What type and length of window would you select if your goal was the measurement of the level of a specific frequency to a high degree of accuracy?
4. What type and length of window would you select to identify a frequency in the presence of a very high noise floor?
5. What is the benefit of windows that exhibit constant overlap and add? Under what circumstances would we select these types of windows?
6. What family of windows would you select for a spectrogram, in which reconstruction back to the sample domain is not required?
7. What technique can be used to obtain both good temporal and frequency resolution on a spectrogram?

13.6 Project – automatic feedback control

Consider this common audio problem: sound reinforcement is being used, and the microphone is pointed towards a loudspeaker – it is well known that feedback will be heard, in the form of a high-level, mid- to high-frequency that rings for as long as these conditions are present. In this project, you will design an automatic feedback control algorithm that identifies the feedback frequency and suppresses it using real-time spectral processing.

To complete this lab, you can utilize the code from Programming example 13.4.1. We will be assuming that the FFT bin with the highest magnitude is related to the feedback. For this lab, instead of converting a band-stop filter to the frequency domain, we will instead follow these steps:

- Extract the magnitude and phase components of the frame
- Find the bin with the highest magnitude
- Set the magnitude of this bin (and a few to either side) to 0
- Reconstruct the complex spectrum according to $Y = |X| \cdot e^{j \angle X}$
- Convert back to the sample domain using **ifft()**

Assignment

1. Download an audio file of microphone feedback from DigitalAudioTheory. com
2. Implement a script to identify the peak frequencies in each frame, and zero out its magnitude
3. Include a variable called "width" to increase or decrease the number of bins that are being set to zero. Comment on the width that you find most appropriate.
4. Listen to the cleaned signal – were you able to reduce the different the feedback?
5. Apply the feedback reduction only when the maximum frequency peak is above some settable threshold

- Hints
- To find the bin with the largest magnitude, use **max**() to return not only the maximum value but also the index of the largest frequency peak. Then use that index to zero out the magnitude at the offending frequency.
- When you find the offending frequencies, you will need to also find the bins corresponding to their conjugates. These will be mirrored about the Nyquist frequency.
- Ensure that your bins are integer values by using round().
- Don't forget to apply **real**() following **ifft**() to remove any residual imaginary components.

14

Analog modeling

Chapter 14 deals with topics from analog circuit analysis, such as the *s*-domain and Laplace transform. While a thorough understanding of these principles is not required to grasp the concepts of this chapter, for the inquiring mind, some resources on the fundamentals of these topics are listed in Chapter 1.4.

In this chapter, you will learn about two methods of converting an analog IR or transfer function to a digital filter, with matching characteristics. However, each of these methods has its own drawbacks, which will also be examined in detail. While these methods are good when an analog system (known as a *prototype*) is well-described, sometimes we may want to derive a digital model when only the magnitude response is given. For these situations, we will, instead, formulate a phase response directly from the magnitude response, creating a stable digital filter that matches the specified magnitude response.

Finally, while the vast majority of this text covers linear audio systems, in Chapter 14, we will also revisit the ESS from Chapter 11.3 as a way to also characterize distortions introduced by certain types of non-linear audio DUTs, such as diodes, triodes, tubes, and transistors.

14.1 Derivation of the *z*-transform

In this section the relationship between the Laplace transform and the *z*-transform will be examined. Let's start by recalling the definition of the Laplace transform, which is given by

$$X_c\left(e^{j\omega}\right) \overset{def}{=} \int_0^\infty x_c(t) \cdot e^{-j\omega t} \cdot dt \tag{14.1}$$

The Laplace transform allows us to think about a signal as a sum of increasing moments of the signal; however, it is conceptionally similar to think of this as a complex frequency analyzer. Note the two continuous variables, t and ω – for any instantaneous complex frequency $e^{(j\omega)}$, every moment of time, t, of the signal, x_c contributes. From the Laplace transform, if we replace the actual integral with the definition of the definite integral, then we obtain

$$X_c\left(e^{j\omega}\right) = \lim_{\Delta t \to 0} \sum_{n=0}^{\infty} \left[\frac{x_c(n \cdot \Delta t)}{\Delta t} \right] \cdot e^{-j\omega n \Delta t} \cdot \Delta t \tag{14.2}$$

Now instead of a continuous variable t, we have an infinitesimal, Δt, which is incremented by n. Next, if we, instead, don't consider t approaching 0, but rather stopping at T_S, then we can make this variable substitution and drop the limit function.

$$X_c\left(e^{j\omega}\right) = \sum_{n=0}^{\infty} x_c(n \cdot T_s) \cdot e^{-j\omega n T_s} \tag{14.3}$$

Recall that we previously defined $x[n]$ as the sampled version of $x_c(t)$ with a sample period of T_s. So, we can drop the T_s, since it is implied in our digital sequences, obtaining

$$X_c\left(e^{j\omega}\right) = \sum_{n=0}^{\infty} x[n] \cdot e^{-j\omega n} \tag{14.4}$$

Last, recall that we defined $z = e^{j\omega}$ giving

$$X(z) = \sum_{n=0}^{\infty} x[n] \cdot z^{-n} \tag{14.5}$$

What this tells us is that the z-transform is actually the digital equivalent of the Laplace transform, where we are using sampled time instead of continuous time. The relationship mapping s to z is given by

$$z = e^{s \cdot T_S} \tag{14.6}$$

14. Analog modeling

14.2 Impulse invariance

Impulse invariance is the method of characterizing an analog system or DUT by digitally sampling its IR to obtain $h[n]$ from $h_c(t)$. When we sample an analog signal, the spectrum becomes symmetrically bandlimited (at f_N) and the spectrum repeats itself every f_S. Therefore, any frequencies of $h_c(t)$ that extend $> f_N$ will result in frequency aliasing. Aliasing can be avoided by filtering the analog IR, or by selecting a high sampling rate to fully capture the HFs of the IR.

14.2.1 Example: RC series filter

Let's consider a series resistor/capacitor (RC) filter, in which we consider the voltage across the capacitor. In this configuration, the capacitor shunts the HFs to ground, leaving an LPF. Without diving into circuit theory, it can be stated, simply by looking up in a textbook, that the IR for such a filter is defined as

$$h_c(t) = \frac{1}{RC} e^{-\frac{t}{RC}}$$

The product, RC, is known as the *time constant*, and it defines the cutoff frequency, as well as the decay of the IR. The corresponding frequency response (again, by textbook lookup) is

$$H_c(s) = \frac{\dfrac{1}{RC}}{s + \dfrac{1}{RC}}$$

Using the impulse invariance method, the IR can be sampled at intervals of T_S, such that

$$h[n] = h_c(n \cdot T_S) = \frac{1}{RC} e^{-\frac{n \cdot T_S}{RC}}$$

Then, we apply the z-transform to the sampled IR, giving

$$H(z) = \sum_{n=0}^{\infty} \frac{1}{RC} e^{-\frac{n \cdot T_S}{RC}} \cdot z^{-n}$$

$$= \frac{1}{RC} \sum_{n=0}^{\infty} \left(e^{-\frac{T_S}{RC}} \right)^n \cdot z^{-n}$$

Solving the summation requires application of power series from Calculus, which reduces to

$$H(z) = \frac{\frac{1}{RC}}{1 - e^{-\frac{T_S}{RC}} \cdot z^{-1}}$$

The next step is to determine the exact gain of H_c at DC. We select DC since this is an LPF; we would select a representative frequency in the passband for other filter types against which we normalize the digital filter. Plugging-in $s = 0$, we see that the gain of H_c at DC is

$$H_c(0) = \frac{\frac{1}{RC}}{0 + \frac{1}{RC}} = 1$$

Therefore, we must set the digital filter, $H(z)$, to also be equal to 1 at DC. On the complex plane, 0 Hz corresponds with $z = 1$. We will solve for a gain factor, G, under these conditions, yielding

$$H(z)\big|_{z=1} = G \cdot \frac{\frac{1}{RC}}{1 - e^{-\frac{T_S}{RC}} \cdot (1)^{-1}} = 1$$

$$G = RC \cdot \left(1 - e^{-\frac{T_S}{RC}}\right)$$

The final digital transfer function, using the impulse invariance method, is

$$H(z) = \frac{1 - e^{-\frac{T_S}{RC}}}{1 - e^{-\frac{T_S}{RC}} \cdot z^{-1}} = \frac{1 - a}{1 - a \cdot z^{-1}}$$

Where $a = e^{-\frac{T_S}{RC}}$. The difference equation can be found by replacing $H(z)$ with $Y(z)/X(z)$, multiplying it by the denominators of both sides of the equation, then taking the inverse z-transform:

$$Y(z) \cdot (1 - a \cdot z^{-1}) = (1 - a) \cdot X(z)$$

$$y[n] - a \cdot y[n-1] = (1 - a) \cdot x[n]$$

$$y[n] = (1 - a) \cdot x[n] + a \cdot y[n-1]$$

14.2.2 Programming example: RC filter

We can simultaneously plot the frequency responses of $H_c(s)$ and $H(z)$ to visualize the artifacts that arise when converting from the s-domain to the z-domain. The cutoff frequency is calculable as

$$f_c = \frac{1}{2\pi RC}$$

If we let $C = 10$ nF and $R = 4.7$ kΩ, then we expect the cutoff frequency to be 3.4 kHz, with the gain at DC to be 0 dB. Although not shown in this example, MATLAB® and Octave both contain a function **impinvar()**, that takes the numerator and denominator coefficients for an analog filter, and returns the comparable coefficients for a digital filter, using the impulse invariance method.

```
fs=48000;
f=0:fs/2;
R=4.7e3;
C=10e-9;

% z transform
a=exp(-1/(R*C*fs));
[H, W] = freqz([1-a], [1 -a], 2*pi*f/fs);

% Laplace transform
Hc = freqs([1/(R*C)], [1 1/(R*C)], 2*pi*f);

semilogx(W/pi*fs/2, 20*log10(abs(Hc)));
hold on;
semilogx(W/pi*fs/2, 20*log10(abs(H)), '--');
```

Since the response of H_c extends beyond f_N some aliasing occurs, resulting in a digital artifact appearing as a deviation from the prototype in the HFs. This is the primary drawback of the impulse invariance method (Figure 14.1).

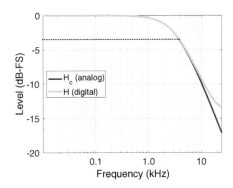

Figure 14.1

An RC-series LPF (black) with a cutoff frequency around 3.3 kHz and its digital model (gray) using the impulse invariance method. While passband and cutoff characteristics match, the digital model diverges from the analog prototype near f_S, due to aliasing contributions.

14.3 Bilinear transformation

The *bilinear transformation* (BLT) is another means of converting an analog transfer function to the digital domain. Given some analog prototype, the BLT maps the $j\omega$ axis of the s-domain to the unit circle, $e^{j\omega}$, of the z-domain, or complex plane. Everything on the left half of the s-plane, from 0 to $-\infty$, maps to the interior of the unit circle on the z-plane, while everything on the right half of the s-plane maps to the exterior of the unit circle on the z-plane, as shown in Figure 14.2.

The BLT preserves stability, phase shift, and maps every point in the analog frequency response, $H_c(s)$ to <u>some</u> frequency in the digital frequency response, $H(z)$. However, while every filter feature or characteristic in $H_c(s)$ is preserved, the exact frequency location is <u>not</u> necessarily preserved. This is especially problematic in the HF, a phenomenon known as *frequency warping*.

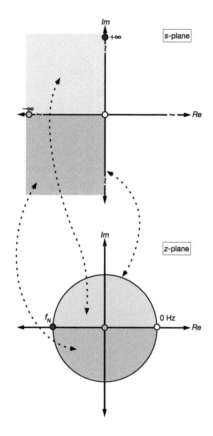

Figure 14.2

Under the BLT, the frequency axis on the s-plane maps to the unit circle on the z-plane, with $+\infty$ Hz and $-\infty$ Hz mapping to $+f_N$ Hz and $-f_N$ Hz, respectively. Everything on the left half of the s-plane maps to the interior of the unit circle. Corresponding frequency locations are indicated by matching colored dots (black, gray, and white).

The BLT is straight-forward to prove, and start with the relationship between s and z. Recall that

$$z = e^{s \cdot T_S} = e^{\frac{s}{f_S}} \tag{14.7}$$

This can be split into two parts:

$$= e^{\frac{s}{2f_S}} \cdot e^{\frac{s}{2f_S}} \tag{14.8a}$$

$$= \frac{e^{\frac{s}{2f_S}}}{e^{-\frac{s}{2f_S}}} \tag{14.8b}$$

The exponential can be approximated by the Taylor Series, which states:

$$e^x = \sum_n \frac{x^n}{n!} = 1 + x + \frac{x^2}{2 \cdot 1} + \frac{x^3}{3 \cdot 2 \cdot 1} + \ldots \tag{14.9}$$

From Equation (14.8b), we substitute in $x = s/2f_S$. If we take just the first two terms from the Taylor series expansion, then the first-order approximation for the exponential is

$$e^{\frac{s}{2f_S}} \approx 1 + \frac{s}{2f_S} \tag{14.10}$$

Then, by plugging in Equation (14.10) into (14.8b), it can be seen that z approximately equals:

$$z \approx \frac{1 + \dfrac{s}{2f_S}}{1 - \dfrac{s}{2f_S}} \tag{14.11}$$

Or, solving for s in terms of z:

$$s = 2f_S \frac{z-1}{z+1} \tag{14.12}$$

Then the digital transfer function, $H(z)$ is found by replacing s in $H_c(s)$, as in Equation (14.12):

$$H(z) = H_c(s), \; s \leftarrow 2f_S \frac{z-1}{z+1} \tag{14.13}$$

14.3.1 Example: RC series filter

Consider again an analog LPF created by a series RC circuit, as in Chapter 14.2.1. If we multiply the transfer function by RC/RC, then it can be rewritten as

$$H_c(s) = \frac{1}{1 + RCs}$$

We can apply the BLT to convert this to a digital filter, by making the substitution, as in Equation (14.13):

$$H(z) = \frac{1}{1 + RC\left(2f_s \dfrac{z-1}{z+1}\right)}$$

$$= \frac{1+z}{(1 - 2f_sRC) + (1 + 2f_sRC)z}$$

$$= \frac{1+z^{-1}}{(1 + 2f_sRC) + (1 - 2f_sRC)z^{-1}}$$

The difference equation of such a digital filter is then given by

$$Y(z)\cdot\overbrace{(1 + 2f_sRC)}^{b_0} + Y(z)\cdot\overbrace{(1 - 2f_sRC)}^{b_1}z^{-1} = X(z)\cdot\left(1 + z^{-1}\right)$$

$$b_0\,y[n] + b_1\,y[n-1] = x[n] + x[n-1]$$

$$y[n] = \frac{1}{b_0}\cdot\left(x[n] + x[n-1] - b_1\,y[n-1]\right)$$

Where $b_0 = 1 + 2f_sRC$ and $b_1 = 1 - 2f_sRC$. The block diagram for this filter is shown in Figure 14.3.

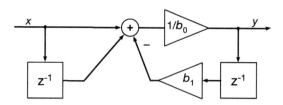

Figure 14.3

The block diagram for a digital model of an RC low-pass filter.

14. Analog modeling

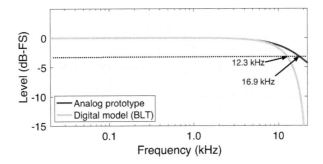

Figure 14.4

Using the BLT, HFs become "squished" to fit within f_N. While this prevents aliasing, it introduces frequency warping. Here, the cutoff frequency of the digital model (gray) does not match that of the analog prototype (black).

If we let C = 2nF and R = 4.7 kΩ, then the denominator coefficients are

$$b_0 = 1 + 2f_s RC = 1 + 2 \cdot 44100 \cdot 4700 \cdot 2 \cdot 10^{-9} = 1.8291$$

$$b_1 = 1 - 2f_s RC = 1 - 2 \cdot 44100 \cdot 4700 \cdot 2 \cdot 10^{-9} = 0.1709$$

Plotting the magnitude response of these coefficients (as in Figure 14.4), we see a filter cutoff frequency of 12.3 kHz.

If we refer back to the equation to predict the cutoff frequency, we see that it should be at

$$f_c = \frac{1}{2\pi RC} = \frac{1}{2\pi \cdot 4700 \cdot 2 \cdot 10^{-9}} = 16.9 \text{ kHz}$$

14.3.2 Frequency pre-warping

This discrepancy between the specified and achieved cutoff illustrates frequency warping inherent with the BLT. Frequency warping occurs because, unlike the impulse invariance method, which directly maps an analog frequency to the same digital frequency, the BLT maps $-\infty$ to $-f_N$ and $+\infty$ to $+f_N$. Whereas the impulse invariance method suffers from aliasing for frequencies outside baseband, the BLT cannot result in aliasing. But the drawback is that warping occurs, especially towards HFs, in which the desired filter characteristic appears at a lower frequency than specified by the prototype.

Fortunately, the warping frequency is predictable, according to

$$f_{warp} = \frac{f_s}{\pi} \text{atan}\left(\frac{\pi f_p}{f_s}\right) \tag{14.14}$$

Where f_p is the frequency specified in the analog prototype, and f_{warp} is the frequency location in the digital domain. From the previous example, this equation exactly predicts the digital cutoff of 12.3 kHz.

It is possible to solve for f_p in terms of f_{warp}; if we specify the where we want the <u>warped</u> frequency to end up, then what would the corresponding <u>prototype</u> frequency need to be? The new prototype frequency, f_p' is determined according to

$$f_p' = \frac{f_s}{\pi} \tan\left(\frac{\pi f_{warp}}{f_s}\right) \tag{14.15}$$

Pre-warping is the process of adjusting the prototype filter to account for frequency warping inherent in the BLT.

14.3.3 Example: pre-warping

If we use the values from Example 14.3.1, then we see that to achieve a cutoff on the digital model of 16.9 kHz, then the prototype's cutoff must be designed to be at $f_p' = 36.8\ kHz$, which can be achieved by keeping the capacitor value the same, but changing the resistor value to $R' = 2.164\ k\Omega$. Using pre-warping, the new denominator coefficients become

$$b_0' = 1 + 2 f_s RC = 1 + 2 \cdot 44100 \cdot 4700 \cdot 2 \cdot 10^{-9} = 1.8291$$

$$b_1' = 1 - 2 f_s RC = 1 - 2 \cdot 44100 \cdot 4700 \cdot 2 \cdot 10^{-9} = 0.1709$$

Plotting the magnitude response of these new coefficients, we see a filter cutoff frequency of 16.9 kHz, as shown in Figure 14.5.

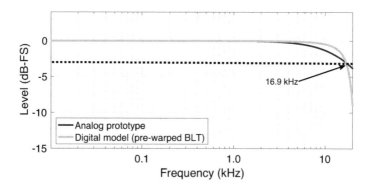

Figure 14.5

By "pre-warping" the cutoff frequency of the analog prototype, the analog and digital models are guaranteed to agree at this one frequency, at least. But, characteristic to the BLT, the digital model's magnitude response hits $-\infty$ dB at fN, while the analog prototype is 4.3 dB at this frequency.

14. Analog modeling

14.4 Frequency sampling

Frequency sampling is the method of capturing the magnitude of the desired frequency response and constructing a digital filter by assuming either linear phase or minimum phase. The IR from this digital filter is obtained by performing the IDFT, producing FIR filter coefficients. Care must be taken since frequency sampling can result in time aliasing (in which the IR tail "wraps around" to the beginning of the IR), an artifact that can be minimized by using a sufficiently large number of frequency samples, N, which will be equivalent to the resulting filter order.

When designing the sampled frequency response, while the magnitude, $A[k]$, is specified by the analog prototype, the phase must be selected as either linear or minimum phase. The phase, $\theta [k]$, will be "constructed" (according to one of these two types), and combined with the magnitude, according to:

$$H[k]= A[k]\cdot e^{-j2\pi\theta[k]} \tag{14.16}$$

Then the IR, $h[n]$ is given by:

$$h[n]=\text{IDFT}\{H[k]\} \tag{14.17}$$

14.4.1 Linear phase frequency sampling

A linear phase filter will have no phase distortion, but a relatively large group delay of $N/(2f_S)$ seconds. The linear phase is constructed by determining the group delay (simply the filter order divided by 2), that then becomes the slope of the phase, which is (of course) linear and intersects the origin such that the phase is 0 rad at frequency bin $k = 0$. The slope of the phase is $N/2$, so at the frequency bin corresponding with the Nyquist frequency, $f_S/2$, must have a phase of $N/4$, with the phase response given as

$$\theta[k]=-\frac{k}{2}, \text{ for } k=0...\frac{N}{2} \tag{14.18}$$

Note that we are constructing only <u>half</u> of the phase response, from the 0 Hz bin to the Nyquist bin. Recall that the spectrum is (nearly) symmetrical about f_N. I say "nearly" for two reasons. The first is because 0 Hz is only represented in the upper half of the unit circle – exactly f_S Hz is equivalent to 0 Hz on the subsequent digital image. The second is because the frequency bins from $k = N/2+1 \rightarrow N-1$ are <u>conjugates</u> of the bins from $N/2-1 \rightarrow 0$, as shown in Figure 14.6.

14.4.2 Programming example: RLC with linear phase

Let's consider a second-order resistor/inductor/capacitor (RLC) band-pass filter, with $R = 10\ \Omega$, $L = 2$ mH, and $C = 5\ \mu$F, which has a center frequency of 1.6 kHz and a bandwidth of 0.8 kHz, given by the following transfer function:

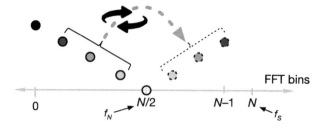

$$0 \qquad \underset{f_N}{N/2} \qquad N{-}1 \quad \underset{f_S}{N}$$

Figure 14.6

It is enough to specify the magnitude and phase up to bin *N/2*, as all bins above that are conjugate symmetric with the bins below *N/2*.

$$H_c(s) = \frac{1}{RC} \left(\frac{s}{s^2 + \dfrac{s}{RC} + \dfrac{1}{LC}} \right)$$

The digital model will have a sample rate of 44.1 kHz, and a filter order of 1024 taps, and will be sampled from $|H_c|$ over the frequency range of 0 to f_S Hz in steps of f_S/N.

```
% analog prototype
R = 10;
L = 2e-3;
C = 5e-6;
f=0:Fs/N:Fs;              %  0 to Fs
Hc = freqs([1/(R*C) 0],[1 1/(R*C) 1/(L*C)], 2*pi*f);

% digital model
Fs = 44100;               % sample rate
N = 1024;                 % length of FFT
```

Next, the linear phase will be constructed. Since the slope is $-N/2$, then at f_N (which is at $N/2+1$ since 0 Hz is at element [1]) the phase is $-N/4$. The baseband transfer function is constructed by combining the magnitude response of the analog prototype with the linear phase response.

```
ph=linspace(0,-N/4,N/2+1); % line from 0 (@ 0 Hz) to -N/4 (@ fN Hz
Hbb = abs(Hc(1:N/2+1)).*exp(j*2*pi*ph); % mag*e^(j*2*pi*ph)
```

Next, the conjugate portion (on the lower half of the complex plane) must be constructed. This is done by taking the conjugate of all bins between 0 Hz and f_N Hz. The digital model, $H_m(z)$, is generated by mirroring the conjugate portion about f_N. The IR is the IDFT (using the **ifft()** command) of Hm

```
Hconj = conj(Hbb(2:end-1));  % take conjugate (excl fN and DC)
Hm = [Hbb fliplr(Hconj)];    % mirror about Nyquist (see Fig. 14-6)

ir = real(ifft(Hm));         % impulse response
```

14. Analog modeling

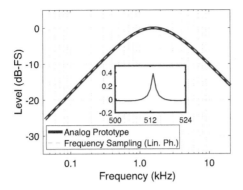

Figure 14.7

The frequency sampling method matches the magnitude response of the analog prototype. The IR (inset) is centered at N/2 + 1 since it is a linear phase.

To verify the IR, a new digital model is created using the IR as numerator coefficients of an FIR filter and evaluated at the same frequency locations as H_c (up to f_N), as shown in Figure 14.7.

```
F=f(1:N/2+1);
H=freqz(ir,1,2*pi*F/Fs);% actual frequency response of IR

% plot
semilogx(F, 20*log10(abs(Hc(1:N/2+1)))); hold on;
semilogx(F, 20*log10(abs(H(1:N/2+1))), '.'); grid on;
figure; plot(ir)
```

14.4.3 Minimum phase frequency sampling

The lowest group delay can be achieved with a minimum phase filter. Minimum phase filters have some useful properties. One such property is that all of the zero locations on its p/z plot are entirely inside the unit circle and, as a result, minimum phase filters are the only type of FIR filter that is invertible. A second useful property that is more material to the problem at hand is that the phase of a minimum phase filter can be derived from the magnitude response! This remarkable property is defined according to

$$\theta_{min} = -\text{Im}\left\{\mathcal{H}\left(\log\left(|H_c|\right)\right)\right\} \tag{14.19}$$

This states that the minimum phase of a filter is equal to the negative imaginary part of the *Hilbert transform* of the log of the magnitude of H_c. The Hilbert transform is a phase shift of a signal of $\pi/2$ (or 90 deg), and interestingly, the z-transform of a Hilbert transformed signal has only zero valued components in the lower half of the unit circle. This tells us that because there are no conjugate pairs in the z-domain, the underlying function must be complex – and indeed, the output of the Hilbert transform (also called the *analytic signal*) is complex.

14.4.4 Programming example: RLC with minimum phase

We will construct a minimum phase IR using the same analog prototype as in Programming example 14.4.2, except, in this case, the frequency range must be defined over $-f_S/2$ to $+f_S/2$, since the **hilbert**() function requires the 0 Hz component to be in the middle (as opposed to the left-hand edge). Next, we construct the phase using the Hilbert transform, according to Equation (14.19).

```
f=linspace(-Fs/2,Fs/2,N); % Hilbert() wants DC in middle
Hc = freqs([1/(R*C) 0],[1 1/(R*C) 1/(L*C)], 2*pi*f);

min_ph = unwrap(-imag(hilbert(log(abs(Hc))))); 
```

Then, the digital model is constructed from the magnitude of the analog prototype and the minimum phase response. The spectrum will be shifted so that the conjugate part is between f_N and f_S, instead of from $-f_N$ to 0 before passing to the **ifft**(), resulting in the IR (Figure 14.8).

```
Hm = abs(Hc).*exp(j*min_ph); % complex freq resp (base band)

% re-structure Hm to go from 0 to Fs, rather than -Fs/2 to Fs/2
Hm=[Hm(end/2:end) Hm(1:end/2-1)];

ir = real(ifft(Hm));         % impulse response
```

Finally, we take the FFT of the IR, analyzing at the same frequency points as the analog prototype – note, baseband is in the second half of the frequency vector.

```
F=f(N/2+1:end);
H=freqz(ir,1, 2*pi*F/Fs);        % actual frequency response of IR

semilogx(F, 20*log10(abs(Hc(N/2+1:end)))); hold on;
semilogx(F, 20*log10(abs(H(1:N/2))), '.'); grid on;
figure; plot(ir)
```

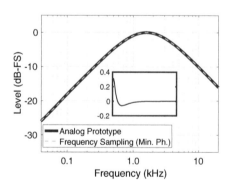

Figure 14.8

Frequency sampling (gray, dashed) of an analog filter (black, solid) using minimum phase response results in an IR (inset) that has its energy compacted near the first sample.

14.5 Non-linear modeling with ESS

The ESS and deconvolution method was discussed already in Chapter 11.3 for capturing linear IRs of an audio DUT. But with the ESS, it is possible to also capture an IR corresponding to the harmonic distortions created by the DUT. This is true for a specific case where the non-linearities are *memoryless*, or instantaneous, and do not depend on previous inputs or future inputs (e.g. "look ahead" on a compressor). Which is to say, if the distortions are arising from hysteresis, coercivity and other zero-crossing distortions, or retentivity, then they are not good candidates for modelling with the ESS method. But distortions generated by waveshaping and amplitude saturation absolutely can be modeled.

It should be clear that linear convolution is not sufficient to fully represent a non-linear DUT. Let's consider an input, x, into a distorting DUT, which has a different IR, h_p, for each harmonic order, p. We could use a form of convolution, if it were modified to consider higher orders of the input, such as the squared and cubed terms (and so on), that each was convolved with their respective IR (h_2, h_3, and so on). A complete, but computationally prohibitive solution is given by:

$$
y[n] = \overbrace{\sum_{k_1=0}^{L-1} h_1[k_1] \cdot x[n-k_1]}^{\text{linear convolution}} + \sum_{k_1=0}^{L-1} \sum_{k_2=0}^{L-1} h_2[k_1,k_2] \cdot x[n-k_1] \cdot x[n-k_2] +
$$

$$
\sum_{k_1=0}^{L-1} \sum_{k_2=0}^{L-1} \sum_{k_3=0}^{L-1} h_3[k_1,k_2,k_3] \cdot x[n-k_1] \cdot x[n-k_2] \cdot x[n-k_3] + \dots
$$

(14.20)

The first-order convolution can be recognized as the first term in this series and successive orders are given by higher-order terms. This approach provides a complete representation of non-linear behavior of a time variant system, provided that enough terms of the expansion are used. It can be seen that the dimension of each term is related to the order of that term. The first-order term is one-dimensional and is plotted as a line. The second term is two-dimensional, and is plotted as a plane, and so on.

Considering just the second-order term, the summation forms a matrix with two axes: the k_1 axis and k_2 axis, which are both time axes corresponding to different time-shift amounts. The diagonal of this matrix represents the instantaneous response, while the lower triangle represents the output to previous inputs, and the upper triangle the output to future inputs. But this model can be simplified if a restriction of memorylessness is placed on the system, whereby $h_p[k_1,k_2,\dots] = 0$ for all instances that $k_1 \neq k_2 \neq\dots$, then Equation (14-20) can be rewritten as

$$y[n] = \sum_{k=0}^{L-1} h_1[k] \cdot x[n-k] + \sum_{k=0}^{L-1} h_2[k] \cdot x^2[n-k] + \sum_{k=0}^{L-1} h_3[k] \cdot x^3[n-k] + \cdots \quad (14.21)$$

Now every term looks more or less like linear convolution, which is by no means blazing fast, but certainly more computationally efficient than the higher-ordered terms in Equation (14.20). The input to the simplified convolution model, given in Equation (14.21), is an ESS signal, which is sinusoidal and depends on the argument $2\pi f n / f_S$, and the argument can be rewritten as $\omega_k n$, yielding

$$y = \sum_{p=1}^{\infty} h_p[n] * x^p[\omega_k n] \quad (14.22)$$

As pointed out by Farina and others [1], when the ESS is processed through a distorting DUT then convolved with the inverse ESS, it produces a sequence of harmonic IRs. For example, if a distortion is produced at 100 Hz, then at the moment that the 100 Hz portion of the ESS passes thru the DUT the linear IR (h_1') is produced, but the h_2' IR is not produced until the 200 Hz portion passes thru, and same for the h_3' IR at 300 Hz.

Importantly, the time difference from 100 to 200 Hz is identical to the time difference from 150 Hz to 300 Hz, and this is also true for every octave span in the ESS. Therefore, all second-order IRs are stimulated simultaneously, resulting in a coherent second-order IR. The same notion applies for all integer harmonics, as shown in Figure 14.9. The harmonic IRs, h_p', that are recovered are

$$y' = \sum_{p=1}^{\infty} h_p'[n] * x[p\omega_k n] \quad (14.23)$$

As opposed to Equation (14.22) where the order, p, is in the exponent of x, in Equation (14.23) the order, p, is now <u>inside the argument</u> of x. The accent on h_p, indicates that this is the <u>harmonic</u> IR following ESS deconvolution.

What we <u>want</u> for a non-linear model are the IRs in Equation (14.22), h_p, but what we <u>get</u> from the ESS method are the harmonic IRs, h_p'. But it is possible to solve for the non-linear IRs in terms of the harmonic IRs. Since the ESS stimulus is a sinusoid, the Power Reduction Theorem can be applied to expand its non-linear terms:

$$\sin^n \theta = \frac{2}{2^n} \sum_{k=0}^{\frac{n-1}{2}} (-1)^{\left(\frac{n-1}{2}-k\right)} \binom{n}{k} \sin\big((n-2k)\theta\big), \text{ for odd } n \quad (14.24a)$$

14. Analog modeling

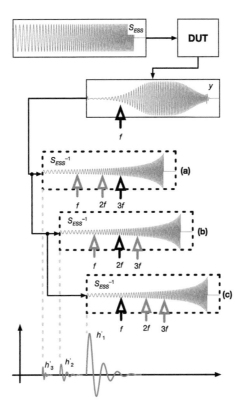

Figure 14.9

An ESS stimulus is processed through a distorting DUT, resulting in y, which is con-volved with the inverse ESS. Three time-shifts of the convolution operation are shown. Any frequency, f, will have harmonics at 2f, 3f, and so on; (a) At a certain shift during the convolution, the 3f frequency of the inverse ESS aligns with the 3f harmonic gener-ated at f, resulting in the 3rd order harmonic distortion, h_3'. (b) At a later point in time, the 2f frequency of s_{ESS}^{-1} aligns with the second harmonic at f, resulting in h_2'. (c) The linear IR occurs when s_{ESS}^{-1} and s_{ESS} are time-aligned, producing h_1'.

$$\sin^n \theta = \frac{1}{2^n}\binom{n}{n/2} + \frac{2}{2^n}\sum_{k=0}^{\frac{n}{2}-1}(-1)^{\left(\frac{n}{2}-k\right)}\binom{n}{k}\cos\big((n-2k)\theta\big), \text{ for even } n \qquad (14.24b)$$

If we consider just the first couple of harmonics, the Equations in (14.24) can be expanded:

$$\sin^2(\omega_k n) = \frac{1}{2} - \frac{1}{2}\cos(2\omega_k n) \qquad (14.25a)$$

$$\sin^3(\omega_k n) = \frac{3}{4}\sin(\omega_k n) - \frac{1}{4}\sin(3\omega_k n) \tag{14.25b}$$

Substituting Equations (14.25) into Equation (14.22), and expanding the series to four terms, we get

$$y = \frac{1}{2}h_2 + \left(h_1 + \frac{3}{4}h_3\right)\sin(\omega_k n) - \frac{1}{2}h_2\cos(2\omega_k n) - \frac{1}{4}h_3\sin(3\omega_k n) \tag{14.26}$$

If the Power Reduction Theorem holds for sinusoids of any frequency, then it should also apply for a sinusoid with all frequencies. Therefore, all of the sinusoids in Equation (14.26) can be replaced with s_{ESS}, after which applying the Fourier transform yields

$$Y = \frac{1}{2}H_2 + \left(H_1 + \frac{3}{4}H_3\right)S_{ESS}(\omega_k) - \frac{j}{2}H_2 S_{ESS}\left(\frac{\omega_k}{2}\right) - \frac{1}{4}H_3 S_{ESS}\left(\frac{\omega_k}{3}\right) \tag{14.27}$$

Where S_{ESS} is the Fourier transform of the s_{ESS}. The cosine functions are represented as simply a shift of a sine to the imaginary axis via a multiplication by the complex constant j. Next the Fourier transform is applied to Equation (14.23), with the series expanded to the first three terms, yielding

$$Y' = H_1' S_{ESS}(\omega_k) + H_2' S_{ESS}\left(\frac{\omega_k}{2}\right) + H_3' S_{ESS}\left(\frac{\omega_k}{3}\right) \tag{14.28}$$

A linear system of equations can be solved by equating the corresponding terms of Equations (14.27) and (14.28), which can be written as the following system of equations:

$$H_{DC} = \frac{1}{2}H_2 \tag{14.29a}$$

$$H_1 = H_1' + 3H_3' \tag{14.29b}$$

$$H_2 = 2jH_2' \tag{14.29c}$$

$$H_3 = -4H_3' \tag{14.29d}$$

Finally, the inverse Fourier transform of the transfer functions in Equations (14.29) gives the harmonic IRs, $h_1...h_3$. This relationship, first described by Farina and others [1], allows for the IRs obtained via ESS deconvolution to be combined to produce a model of harmonic distortion. The time location of the measured IRs, h'_p, is

$$t_p = -\frac{T}{\log\frac{f_2}{f_1}}\log p \qquad (14.30)$$

The location of h_p' depends only on the order, p. Since the IRs are sequentially located, there is a possibility of overlap between successive orders. This is especially true as the order increases as the time difference between IRs compresses. Overlap can be decreased by increasing the ESS duration, or by decreasing the f_2/f_1 ratio. For example, a 1 s ESS from 40 Hz to 20.48 kHz has a second-order harmonic, h'_2, located at $t_2 = -111$ ms and h'_3 $t_3 = -176$ ms. Note that in this example, if h_3' is longer than 65 ms, then it will overlap with h_2'. The generic design of a third-order non-linear model, is described in Figure 14.10.

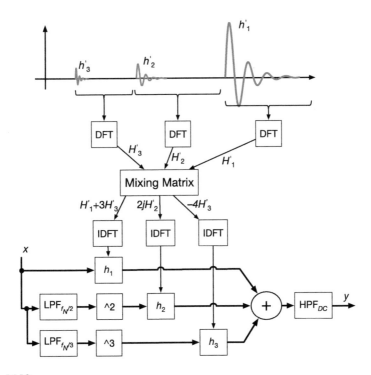

Figure 14.10

The ESS method results in a sequence of IR, corresponding with each harmonic response (up to third order shown here) of the DUT. These responses are converted to the frequency domain, and then mixed according to Equations (14.29), then converted back to the time domain for convolution with various orders (here, up to third order) of the input.

The digital non-linear model includes LPFs on the input prior to raising the order of the input by squaring or cubing. The cutoff frequency is equal to f_N/p, which is done to prevent aliasing, since the following block causes the frequencies to increase by the order indicated. Additionally, following summation of all of the harmonic branches is an HPF that blocks DC. This is in place to reduce the contribution to DC caused by the even orders, as described in Equation (14.29a).

14.6 Challenges

1. A RC high-pass filter is given by the following IR:

$$h_c(t) = -\frac{1}{RC}e^{-\frac{t}{RC}}$$

   a. What is the sampled IR, h[n] using the impulse invariance method?
   b. What is the digital transfer function, H(z)?
   c. What is the gain at the Nyquist frequency, fN?
2. The same HPF has a transfer function of

$$H_c(s) = \frac{RCs}{RCs+1}$$

   a. What is the digital transfer function, H(z) using the bilinear transform?
   b. What is the difference equation, with y[n] on the left-hand side of the equation?
   c. Is pre-warping required for an HPF? Why or why not?
3. Use the system of Equations given in (14.27) and (14.28) to solve for the Equations (14.29).
4. In this chapter, the digital model was solved up to the third order. Expand the solution to the fourth order.
   a. Add another equation (what would be c.) to (14.25) that solves for sin4().
   b. Add another term to Equation (14.26) to include h4.
   c. Add another term to Equation (14.27) to include H4.
   d. Add another term to Equation (14.28) to include .
   e. Update all of the equations in (14.29), and add another equation (what would be e.) to include .
5. Determine the locations of the harmonic IRs, up to four terms, for an ESS of duration 1 second, with an f_1 of 62.5 Hz and an f_2 of 16 kHz.

Bibliography

[1] Farina, A., Bellini, A., Armelloni, E. "Non-linear Convolution: A New Approach for the Auralization of Distorting Systems." Presented at the 110th Convention of the Audio Engineering Society, Amsterdam, 12–15 May 2001. Paper 5359 (4 pp).

Index

T - #0083 - 111024 - C0 - 234/156/12 [14] - CB - 9780367276553 - Matt Lamination